ORIGINS

Creation or Evolution

Richard B. Bliss, Ed.D.

Director of Curriculum Development
Institute for Creation Research

Consulting Editor:

David W. Unfred, M.S., M.B.A., Creation Life Publishers, Inc.

Technical Advisors:

Dr. Gerald E. Aardsma, Ph.D., Nuclear Physics
Dr. Kenneth B. Cumming, Ph.D., Ecologist
Dr. Duane T. Gish, Ph.D., Biochemist

Institute for Creation Research
P.O. Box 2667 • El Cajon • CA 92021

Origins: Creation or Evolution

Copyright © 1988

Institute for Creation Research
P.O. Box 2667, El Cajon, CA 92021
619/448-0900 • www.icr.org

First printing, November 1988
Second printing, September 1989
Third printing, March 1991
Fourth printing, July 1994
Fifth printing, January 1995
Sixth printing, August 1996
Seventh printing, October 2002

ISBN: 0-89051-132-2

Cover: Dave Anderson, Mark Dinsmore, and Jay Wegter

Typesetting and design layout: Gloria Clanin

Artists: Dave Anderson, Jay Wegter, Tim Revenna, Ted Hansen, Marvin Ross, Chris Roth, and Joe Austin

Photo credits: Bill Hoesch (Grand Canyon, p. 33); UPI (Horses, p. 47)

Printed in the United States of America

ACKNOWLEDGMENTS

I wish to express my deep gratitude to the many science teachers and students for their valuable constructive review of the Origin modules, their participation in the piloting process of the original manuscript and, above all, their enthusiasm and encouragement in this truly open and scientific approach to a controversial topic.

Richard Bliss
August, 1988

ABOUT THE AUTHOR

Richard B. Bliss, Ed.D., had more than 39 years experience in all areas of science education. In addition to having taught chemistry, physics, biology, and general science at the high school level, his was adjunct professor, teaching science methods to teachers in the University of Wisconsin System. He was engaged in biological research and obtained several National Science foundation grants and fellowships during his academic career. He developed a hands-on curriculum in science for K–6 elementary students that was based on the most current research in science education. Dr. Bliss was a frequent speaker on the creation/evolution issue in the U.S. and other countries. He is now deceased.

TO THE STUDENT

This book deals with the most exciting unknown in science. This mystery has stirred the thinking of thoughtful men over the centuries and continues to do so today. Many scientists are searching for discoveries that will explain how life came into existence. Governments are spending billions of dollars to search for life on other planets, yet the answer to this question is still a scientific mystery.

In order to preserve honesty in science the National Academy of Sciences made an open resolution to all scientists in April of 1976 called "An Affirmation of Freedom of Inquiry and Expression." Here is a portion of the resolution:

> . . .That the search for knowledge and understanding of the physical universe and of living things that inhabit it should be conducted under conditions of intellectual freedom, without religious, political or ideological restrictions . . .that freedom of inquiry and dissemination of ideas require that those so engaged be free to search where their inquiry leads . . . without political censorship and without fear of retribution in consequence of unpopularity of their conclusions. Those who challenge existing theories must be protected from retaliatory reactions.

No scientist can truly say that he is free to do science unless he can abide by a resolution such as this. Science and scientific investigation must have these freedoms or the basis of the whole scientific enterprise falls apart. Science is not conducted in a vacuum and must consider information and thought from other areas of knowledge. When the student begins to study the origin of life, there is also much that has to be considered from the historical and religious fields. Science can bring to light certain data about the distant past and history can bring us closer to understanding their meaning.

This book will give you a picture of two scientific viewpoints of origins: creation and evolution. A searching, inquiring mind is a free mind, and we as educators want this for our students. Unlike many textbook presentations of the origins problem, we are not going to withhold any information that will have a significant bearing on this matter. You will read about what is known and the suggested explanation from evolutionists and creationists. I trust that you will have an exciting time exploring the mystery of life's origin.

Table of Contents

What does science say about the origin of life? Looking at the data from origin-of-life experiments, the most reasonable scientific inference one can make is that the first living organism was designed by an Intelligence, not a product of chance and time.

In the information molecule DNA, we see a highly ordered and complex structure that even many evolutionists no longer believe could have come about by accident. The probability that the DNA molecule is the result of chance and time is zero.

When bones, organs, or molecules of different living organisms look similar, does this mean that they are related to each other? According to genetics and a close examination of the data, likeness does not mean relatedness. Homology has given many scientists a false hope as a proof of evolution.

Does variation, mutation, migration, isolation, and natural selection produce new, more complex life forms? The answer is no. There are natural limits or barriers to change that conform to the creationist concept of stasis.

Where in the world is the standard geologic column? Are index fossils reliable dating tools? The true scientific method exposes faulty reasoning used in some scientific circles.

What the fossil record does record is that catastrophes have devastated Earth in the past, resulting in the rapid burial of many plants and animals. What best explains such widespread destruction? The best model is a worldwide flood catastrophe.

Although the fossil record has disappointed evolutionists, there have been rare finds that have been proclaimed as the missing transitional organisms needed by evolution. But when these missing links of evolution have been closely examined, they were found to be still missing.

Some scientists have been so eager to find the evolutionary ancestor of humans that every fossil they discover is proclaimed to be the "missing link." This non-scientific behavior has led to the acceptance of many hoaxes and frauds by the scientific community, as well as by the public.

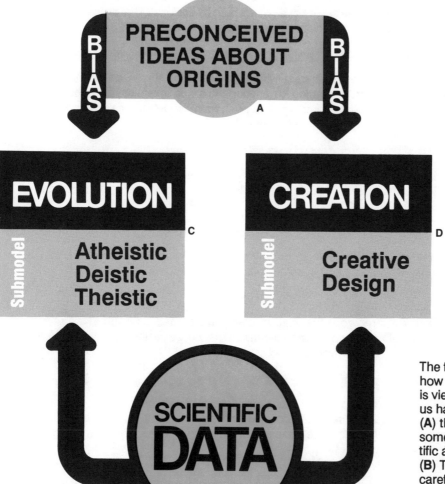

The figure shows how the subject of origins is viewed differently. All of us have preconceived ideas (**A**) that are the result of some bias; this is non-scientific and results in dogma. (**B**) To be scientific we must carefully look at the data and try to find the model that fits it best; i.e., evolution (**C**) or creation (**D**).

INTRODUCTION

Our study into the origin of life begins by briefly examining the nature of science.

Has science alone produced evidence that tells us for certain how the earth was formed? Is there scientific evidence about how life began? What about the solar system and all the things beyond; do we know how these were formed? On the basis of only scientific evidence, the best answer to these questions, and many others, is that no one knows. It is as though we are looking through a tiny keyhole into the night of outer space and then making assumptions about the complexity and variety of the universe. Data gathered about the past serves to inspire the inquiring mind of man. This open frontier of knowledge affects how people view the world about them. When we consider first beginnings, we cannot avoid coming face to face with ultimate causes (the very first responsible cause: accident or an all-powerful God) and ultimate meaning (why life came to be and for what purpose). These issues involve our religious beliefs.

Ken Ham, author and lecturer, has simply and brilliantly shown the relationship between religious beliefs and the interpretation of scientific evidence about origins. He writes:

> It is astonishing in this so-called "scientific age" that so few people know what science really is or how it works. Many think of scientists as unbiased people in white laboratory coats objectively searching for the truth. However, scientists come in two basic forms, male and female, and they are just like you and me. They have beliefs and biases. A bias determines what you do with the evidence, especially the way in which you decide that certain evidence is more relevant or important than other evidence.
>
> Many people misunderstand bias, thinking that some individuals are biased and some are not. Consider an atheist, a person who believes there is no God. Can the atheist entertain the question "Did God create?" As soon as he even allows it as a question he is no longer an atheist...
>
> Atheists, agnostics and theists hold to religious positions; what they do with the evidence will be determined by the assumptions (beliefs) of their religious positions. It is not a matter of whether one is biased or not. It is really a question of which bias is the best to be biased with. (Ham, 1987)

We can't avoid acknowledging our biases. Our beliefs will determine what we do with the evidence. If we can believe that the Scriptures are accurate and that their story of the origin of life is true, then this belief will provide a framework or model for interpretation of the data we are collecting. In the same way, if we believe that life began by accident on earth, or evolved in outer space, we use this belief to establish a different framework of data interpretation. Whatever the first cause, one can readily see that we must apply our religious belief at some point to the evolution or creation question. True science can best deal with those things that can be observed or measured. When we develop our scientific models, they must come from the best scientific data and these data are derived from the skills of scientific inquiry. This has to be the basis for study in science.

Students as well as scientists find themselves choosing a particular model for origins for a variety of reasons. Some times these reasons are not scientific at all. Often we find that scientists can be carried away with their personal viewpoints; they no longer look at the data objectively. One model or the other is assumed to be true without further thought. With this in mind it is our sincere hope that you will look at the origin of this earth and the life upon it with a searching mind. The topic of origins can become very exciting when one compares all available data with his model of origins. An inquiring person will challenge his models. Any scientist, or student, who is searching would want to examine the data, **scientific and historical**, in order to make a good decision.

The two general scientific models of origins are *"creation"* and *"evolution."* There are also **submodels** under these. We must recognize that religion, which is outside

8

of science, also plays an important part. The following short explanation of religious views will help you understand this relationship.

Atheistic evolution submodel — no supernatural being was involved in the process of developing life. All life arose by **naturalistic, mechanistic** processes without any direct purpose or help from any deity or divine force. *This submodel is the one typically presented in public science textbooks.*

Deistic evolution submodel — a deity started the first life, and naturalistic processes allowed life to evolve into what we see today. This submodel is popular among many Hindu and Buddhist scientists, as well as scientists who are at a loss to explain the ultimate cause of matter, energy and space.

Theistic evolution submodel — God not only started the life process, but directed it through all the stages of evolution; the life we see today is the result of this directed purposeful process. Christians who have difficulty believing in the historical accuracy and literal authority of the Bible often find this submodel attractive.

Creative design or abrupt appearance submodel — all life and life processes were designed to function just as we observe them today. Life appeared quickly on the scene. The designer for all of this was the God of creation. The Creator not only created and designed life but also gave purpose to it.

With respect to the **submodels** and the general models, each one depends upon some religious belief or some kind of faith. Nobody was there in the beginning. We can only try to form intelligent ideas about what happened from the data we have collected and the observations we have made.

Origins: Creation or Evolution will not deal with all religious and sectarian views, but will deal with Biblical historicity where it sheds light on origins. In all cases, our purpose will be to bring the controversy into a clearer focus and encourage you to find the model that fits the scientific data best. With the help of this book, you will be able to make these decisions more intelligently. Which model, creation or evolution, best fits the data? No one was there. Ultimately you will have to make your own decisions.

PROCESS SKILLS OF SCIENCE

1. **Observation** — using all your senses
2. **Classification** — placing your observation into the most logical categories
3. **Measuring** — making careful measurements
4. **Inference** — establishing best guesses with the scientific information you have
5. **Prediction** — making predictions based upon available scientific evidence
6. **Interpretation** — making a best interpretation of the scientific data you have available
7. **Experimentation** — performing science experiments to add to your knowledge
8. **Operational Objectives** — establishing objectives that you can work with and are clearly understood
9. **Model Building** — building the best temporary models from the data you now have

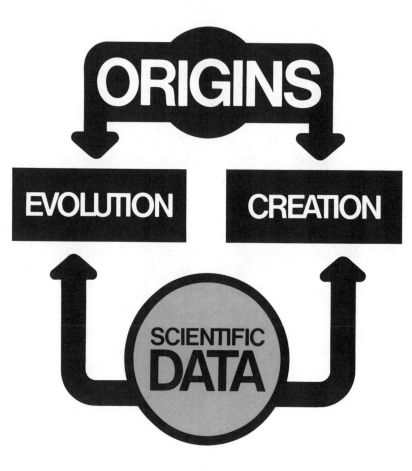

CHAPTER ONE

Life's Beginning — Failed Experiments

What does science actually know about the origin of life? As we examine the ideas put forth by evolutionists—as well as the ideas from scientific creation—we will get a better picture of how little is really known, scientifically, about this topic.

Since no one was there in the beginning, any model for the origin of life requires a great deal of faith when describing first matter or first life. Creationists explain the origin of the first life as appearing on the scene in a very short period of time. Evolutionists insist that life changes so slowly into more complex forms of life that the actual process of evolution cannot be seen.

Two basic questions in this chapter are: Can non-living chemicals ever produce even the simplest living cell by random or chance processes? Secondly, do all living systems require simultaneous symmetry (order), purpose and inter-dependence (all parts of the cell working together at once)?

The Early Earth

The first experiments attempting to create life from non-living chemicals used a methane and ammonia atmosphere. It was believed that such an atmosphere would offer the best environment for the chance occurrence of the first cell. Today many scientists are arguing that there never was a reducing atmosphere in the first

REDUCING

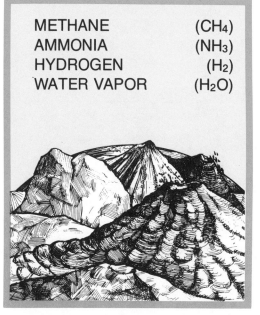

METHANE	(CH_4)
AMMONIA	(NH_3)
HYDROGEN	(H_2)
WATER VAPOR	(H_2O)

OXIDIZING

WATER VAPOR	(H_2O)
OXYGEN	(O_2)
CARBON DIOXIDE	(CO_2)
NITROGEN	(N_2)

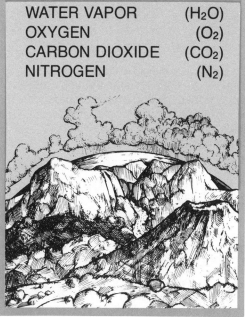

Figure 1 New data show that the earliest atmosphere on earth may have been an oxygen atmosphere.

place. NASA published a report stating that a new model containing oxygen must be used. If oxygen was part of the original atmosphere, then the origin-of-life chemistry using a reducing atmosphere model won't work.

Stanley Miller's Experiments

One of the early origin-of-life experiments utilizing a reducing atmosphere was by Stanley Miller. Examine the drawing of Dr. Miller's apparatus and then follow some of the arguments against his ideas.

A vital part of Miller's apparatus was the **cold trap** for collecting the products as they were formed. If the chemicals that were formed in the apparatus were allowed to circulate back into the system **not** having this cold trap, *they would be destroyed at a much faster rate than they were produced.* Biochemist Miller knew this to be true, so he placed a cold trap in the system. **Would this work in an evolutionary plan?** Where could a system like this be found in the real world? The fact is, no such system has ever been found in nature. Furthermore, if the products are kept isolated from the energy source, no further progress would be possible because every step is more complex and requires more energy.

A. I. Oparin's Experiments

Another scientist who attempted to duplicate origin-of-life chemistry is the Russian chemist A. I. Oparin. He believed that his "coacervate hypothesis" could eventually show the way to producing living cells. "Coacervates" are like fat droplets in a bowl of soup. The idea was that these globules would join together to form larger globules and ultimately a living cell. This was

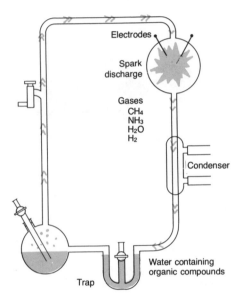

Figure 2 In Stanley Miller's famous apparatus, electric discharge through a special mixture of heated chemicals produced biologically interesting molecules (amino acids, etc.), which were collected and drawn off at the cold trap.

Electrodes

Spark discharge

Gases
CH$_4$
NH$_3$
H$_2$O
H$_2$

Condenser

Water containing organic compounds

Trap

popular for a time until chemists began to realize how unstable coacervates were. They would disassociate — fall apart — after a time. Scientists also observed that anything near the coacervate could be absorbed into it. This fact alone would destroy the chemical orderliness required by a living cell. Chemicals that were destructive would be absorbed into the coacervate just as quickly as the helpful kinds. In other words, there would be no selectivity in the coacervate film (membrane-like). Most scientists seem to agree that the coacervate hypothesis would not support evolutionary expectations for the first spark of life.

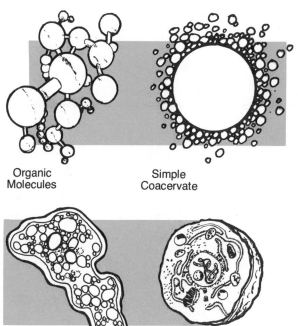

Organic Molecules

Simple Coacervate

Complex Coacervate

First Simple Living Organisms

Figure 3 A. I. Oparin's coacervates seemed to grow from simple to complex.

Sidney Fox's Experiments

Another effort at duplicating the chemical events that might produce life was that of Sidney Fox. Protein-like molecules were produced by heating pure, dry amino acids at 150°-180° C for several hours. He **thought** that something similar to this might have happened eons of years ago on the side of a volcano. When he dissolved the chemical product in hot water, he observed small microspheres forming. He called these "proteinoids." They seemed to bud and grow. This idea was attractive for a very short time until scientists began to ask questions.

Some of the questions were: where could you ever find dry, pure amino acids on the earth? Answer: only in a chemist's laboratory. If the temperature was this hot for over ten hours wouldn't this destroy the amino acids? Answer: yes. Volcanoes are wet places; how could you find a required dry spot for six hours? Answer: it is very unlikely that you would. The Fox hypothesis doesn't fit

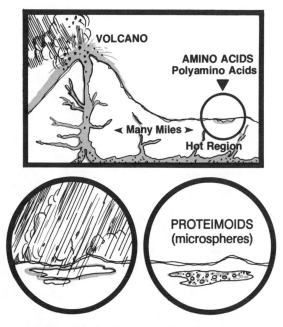

VOLCANO

AMINO ACIDS Polyamino Acids

◄ Many Miles ►

Hot Region

PROTEIMOIDS (microspheres)

Figure 4 Fox's theory requires that amino acids be subjected to a great deal of heat (175°). These amino acids would form polymers; be washed away (between 4-8 hours) by rain; collect in an isolated pool; and form proteinoids which would evolve into living cells.

the chemistry needed for the origin of the first cell (see also Bliss and Parker, 1979).

The scientific creationist position is that the complexity in living systems is far too great and ordered to ever occur through chance processes in nature. While these are interesting experiments, they all require creative design; that is, human intelligence. Creationists believe that an intelligence was necessary to cause the abrupt appearance of living things. While evolutionists are trying to find non-intelligent ways for life to occur, the creationist insists that an intelligent design must have been there in the first place.

Summary

Considering all of the scientific data from the above experiments, we can conclude the following:

1. The Miller experiment assumes a primordial reducing atmosphere (methane and ammonia). There is no evidence that such an atmosphere ever existed. The atmosphere appears to have always contained oxygen and carbon dioxide (an oxidizing atmosphere).

   A very important **laboratory modification**—the cold trap—was placed in the Miller apparatus to keep the reaction going in the "right" direction. Without this trap the energy in the system would have destroyed the product as fast as it formed.

2. A. I. Oparin's coacervates have an unstable characteristic about them. The thin film which surrounds the coacervate is so fragile that it breaks easily, making its use as a model for the origin of the first cell doubtful.

   The coacervate hypothesis breaks down in other significant ways. The coacervate will absorb any molecule near it. This means that harmful chemicals, as well as good ones, will react adversely. This is not an acceptable condition for the first cell.

3. Fox's experiment seems to fail in the very beginning. Where would the pure amino acids come from? This question is basic and unanswered.

   Fox's amino acids would cause restrictions placed upon the chemistry. Dry heat conditions could only last for a few hours, and there would have to be a total absence of moisture during this heating period.

4. The tiniest living organism (such as the bacterium) displays symmetry (order), purpose (reason to be), and interdependence (one part dependent upon the other).

5. The conditions in any origin-of-life experiment were created by human intelligence (chemist Stanley Miller) and therefore cast doubt on whether random chance processes could have produced even the simplest building blocks of life.

The conclusion is simple: to date, scientific experiments that allegedly support an evolutionary hypothesis of life arising from non-living chemicals do not work. The data points toward **symmetry**, **purpose** and **interdependence** caused by intelligent design.

STOP AND THINK

Scientists who want only an evolutionary model for the origin of life believe that the "equation" for life's beginning was

Energy + Matter + Lots of Time = Life.

All that is required for life to begin is energy (light and heat) and matter (a random selection of chemicals) acting together long enough for the impossible to become possible. It would be accurate to say that evolutionary scientists believe that these events occurred in a closed system, that is, a system where the products are trapped within and don't escape.

Creation scientists argue that any valid scientific model for the origin of life has to include intelligent information and this requires an initial blueprint, which itself comes from a designer. Reduced to an equation,

Energy + Matter + Outside Information = Life

If you supply radiation (a form of energy) to a sterile can (closed system) of sardines (matter) for a long (very long) period of time, will a reasonable person expect to get life? How is this simple analogy similar to the experiments of Miller, Oparin and Fox? How is it different?

Are Your Molecules Left or Right-Handed?

Proteins such as those found in hair, finger nails, muscle, skin—in all living tissues—are made from amino acids. The most fascinating fact is that all of these amino acid molecules in living organisms must be **left-handed**. A molecule is called left- or right-handed depending on its chemical arrangement and how that arrangement rotates a special kind of light. Left-handed molecules rotate a plane of *polarized* light to the left, while right-handed molecules rotate the light to the right. For living systems this fact is a matter of life or death!

But the molecules Miller made included not only the left-handed amino acids required in living systems; they also included equal quantities of amino acids that would end life. If even one right-handed amino acid got into an enzyme the enzyme would be biologically useless. In other words, left to time, chance, and their inherent chemical properties, any hope of producing life by this method would be zero.

Another significant part of origin-of-life chemistry is that of the sugars. Sugars are one of the energy sources for living systems and a very important part of the DNA molecule. Well, it just so happens that sugars in living systems rotate polarized light to the right. **They are right-handed**. Just like amino acids, when left alone to reach chemical equilibrium, 50% will rotate to the right and 50% to the left. If a left-handed sugar appeared in the chain it would destroy the biological usefulness of that sugar. The living organism could not get energy from the sugar because it would not fit the enzyme needed to break it down.

This arrangement is just like a lock and key. If the lock and key won't fit, the chemistry doesn't work. All living systems from the very start need both amino acids and sugars; *these amino acids and sugars are highly specific and cannot deviate from the pattern dictated by the information in the DNA model.*

Chemistry, then, is not our ancestor; it's our problem.
(*What Is Creation Science?*, Morris and Parker, 1987)

Figure 5 Left to themselves, amino acids, sugars, and other biological molecules naturally react in destructive ways, and this makes some scientists wonder how such molecules in the early ocean could evolve toward life.

16

CHAPTER TWO

DNA: By Accident Or Design?

This chapter will deal with some of what we know about the DNA molecule. We will look at how the DNA molecule relates to origin-of-life experiments, as well as the probability of DNA beginning by chance. What are the odds that information molecules such as DNA could happen by chance evolutionary processes? This question is basic to the study of how life began.

In all living systems (even some viruses), the most interesting example of molecules working together is how DNA provides the information codes to make proteins. DNA is a unique molecule that uses a three-molecule code (triplicate code). This triple code lines up molecular sequences and organizes the amino acids, the building blocks that make proteins. The DNA acts like a blueprint that actually points the molecules in the correct predetermined direction. A brief description of the DNA molecule shows that it has four building blocks, **guanine (G)**, **cytosine (C)**, **adenine (A)**, and **thymine (T)**, which are called nitrogen bases or purines (A and G) and pyrimidines (C and T).

Triplicate combinations of these molecules in the DNA (or gene) determine the correct order of amino acids required to make each protein. This is done in a complex, efficient, and highly specific manner. For example, the amino acid "proline" is specified by the DNA code C-A-T (Cytosine-Adenine-Thymine). In this manner each of the many amino acids have their own special code. The drawings on the following page will help you to understand this better:

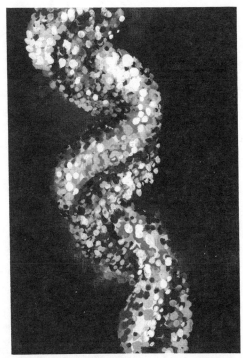

Figure 6 This illustration shows a segment of a DNA molecule.

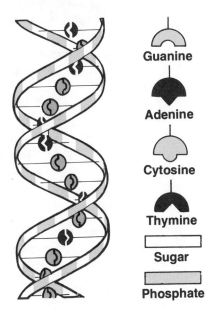

Guanine

Adenine

Cytosine

Thymine

Sugar

Phosphate

Figure 7 The DNA molecule has a precise coding arrangement.

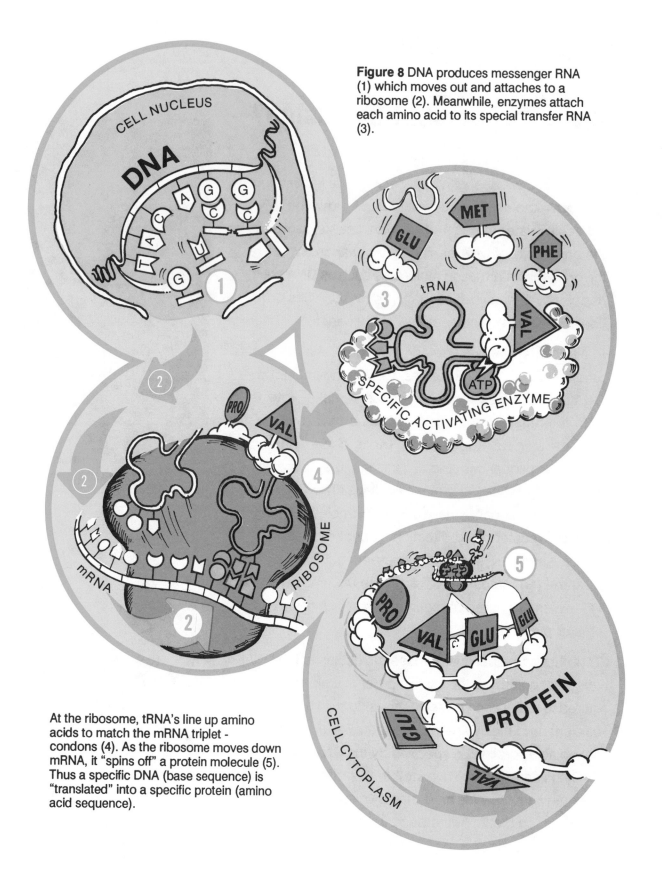

Figure 8 DNA produces messenger RNA (1) which moves out and attaches to a ribosome (2). Meanwhile, enzymes attach each amino acid to its special transfer RNA (3).

At the ribosome, tRNA's line up amino acids to match the mRNA triplet - condons (4). As the ribosome moves down mRNA, it "spins off" a protein molecule (5). Thus a specific DNA (base sequence) is "translated" into a specific protein (amino acid sequence).

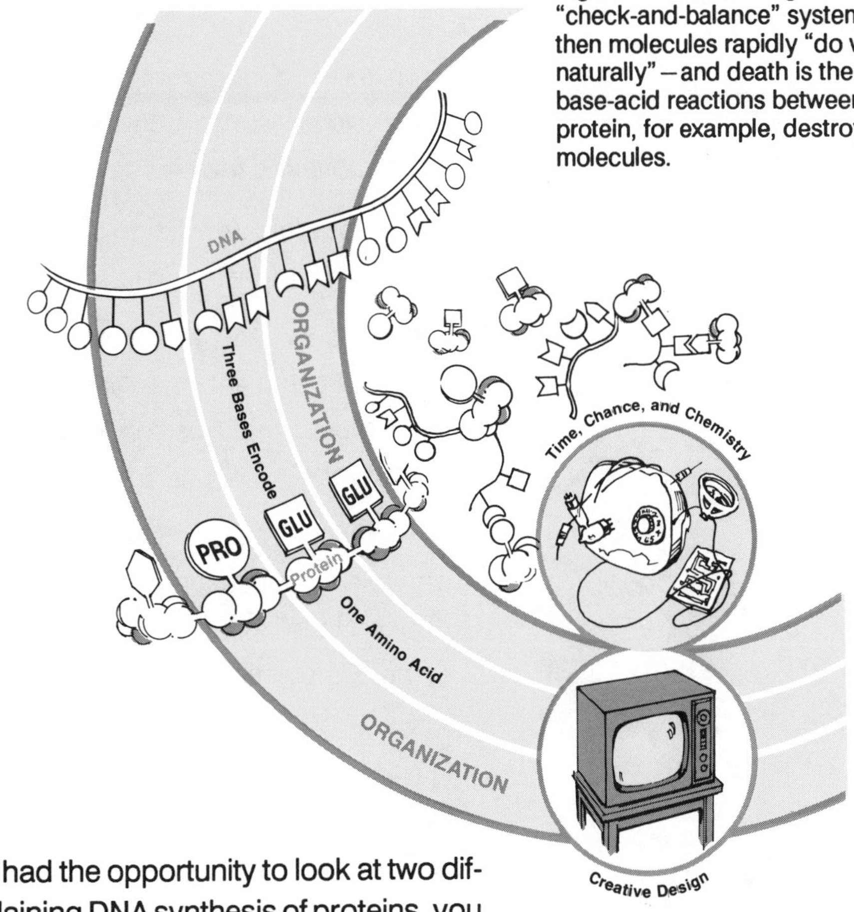

Figure 9 When a living cell's intricate "check-and-balance" system breaks down, then molecules rapidly "do what comes naturally"—and death is the result. Natural base-acid reactions between DNA and protein, for example, destroy both of these molecules.

Now that we have had the opportunity to look at two different models explaining DNA synthesis of proteins, you can better appreciate how complex this whole chemical story becomes. Think about this: in order to make one protein a cell uses over **70 specific proteins** (which had to be made by DNA) and a constant supply of amino acids, energy, and specific enzymes for each of these amino acids (enzymes are also proteins). In addition to this, we find some startling facts. First among these facts is that there are about twenty proteins that must be active in the construction of the DNA molecule. Second, and of equal importance, is that these proteins (enzymes), according to everything we understand, must have been constructed by DNA in the first place. *Now if proteins (enzymes) are coded by DNA, then which came first in an evolutionary process, the protein or the DNA?* Creationists predict that the only logical conclusion is that this system was created, fully functional, by an intelligent designer.

Figure 10 Creative intelligence gave circuit boards and phosphors properties for television picture transmission. They could not develop by themselves. We can conclude for similar reasons that creative intelligence established the code of DNA.

19

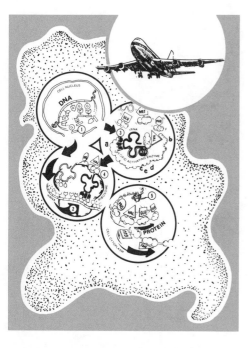

Figure 11 Living cells use over 70 special kinds of protein and RNA molecules to make one protein following DNA's instructions. What we know about airplanes convinces us that their flight is the result of creation and design. What scientists know about the way living cells make protein suggests, just as clearly, that life also is the result of creation.

Two world-famous astronomers and mathematicians, Sir Fred Hoyle and Chandra Wickramasinghe, considered the probability of the spark of life forming by chance, evolutionary processes. The following are some of the discoveries they made:

1. Enzymes, which are vital to all of biology, do not leave even a hint of their origins.

2. There are ten to twenty distinct amino acids in the structure of the enzyme. All of these amino acids must be in the correct position in order for the enzyme to do its work.

3. Consider the chance that a **random array** of twenty amino acids necessary for making up proteins would happen to fall in the correct order for any specific enzyme. Calculations would force the conclusion that this just couldn't happen.

Hoyle and Wickramasinghe concluded that the probability of getting life anywhere in the universe from evolutionary processes was as reasonable as getting a fully operational Boeing 747 jumbo jet from a tornado going through a junkyard!

Considering the complexity of the simplest biological systems and the "impossible" probabilities of an accidental formation of the fundamental molecules of life, it appears that the rational mind must reject the idea that the origin of life could occur by chance evolutionary processes. It appears that scientific objectivity strongly points to an intelligent designer. The intelligence that appears to have been incorporated into the living cell is more complex than scientists ever thought.

ACCIDENT OR DESIGN?

Let's consider some factors that are involved in probability. If you recall, we said that all proteins (remember enzymes are proteins) require left handed amino acids (see page 16). Now, using this fact of life science, let's try using probability theory (Calculations submitted by Dr. Monty Kester).

(Doing the calculations)

I. What is the probability of forming one all left-handed (L), 400-amino-acid protein, from a normal 50% mixture of right handed (D) and left handed (L) forms?

    A. The probability of each (L) linking is 1/2

    B. If we deduct a fair share for <u>glycine</u>, say 20, then the probability would be

$$\frac{1}{2} \times \frac{1}{2} \times \ldots 380 \text{ times } (400 - 20) = \frac{1}{2^{380}} = \cdots \frac{1}{10^{114}}$$

(incidentally, you must convert to logarithms)

II. What is the probability of 124 such proteins, the number needed for the simplest possible self-replicating system forming?

$$\frac{1}{10^{114}} \quad \text{(probability of one protein forming)}$$

$$\frac{1}{10^{114}} \times \frac{1}{10^{114}} \ldots 124 \text{ times } = \frac{1}{(10^{114})^{124}} = \frac{1}{10^{14,136}}$$

III. What is the probability of 124 all (L), 400-amino-acid proteins forming, if there is a 99% surety that (L) will preferentially link to (L)?

    A. The probability of each (L) left handed amino acid linking is

$$\cdots \frac{99}{100}$$

    B. The probability of 380 (400-20 glycine) L - amino acid linking in succession is:

$$.99 \times .99 \times .99 \ldots 380 \text{ times } = .99^{380} = \frac{1}{10^{1.7}}$$

    C. The probability of 124 proteins forming is:

$$\frac{1}{10^{1.7}} \times \frac{1}{10^{1.7}} \ldots 124 \text{ times } = \frac{1}{(10^{1.7})^{124}} = \cdots \frac{1}{10^{210}}$$

When we consider the facts related to these numbers we can easily see that the random chance of evolutionary theory will not likely be useful. Consider what those numbers mean in general comparisons:

        1. Age of the universe (evolutionary assumption) in seconds = 10^{18}

        2. Diameter of the universe in inches = 10^{28}

        3. Diameter of the universe in A° (angstroms) = 4×10^{37}

        4. Mass of the Milky Way in grams = 3×10^{44}

        5. Number of atoms in the universe = 5×10^{78}

Physicists conclude that events whose probabilities are extremely small **never occur** (1 chance in 10^{15}).

Think about this: a scientist can sequence (put together) 124 proteins in a matter of hours. This shows the need for intelligence behind design.

Can Aluminum Fly?

I'm sure that sounds like a trick question. By itself, of course, aluminum can't fly. Aluminum ore in rock just sits there. A volcano may throw it, but it doesn't fly. If you pour gasoline on it does it fly? Pour a little rubber on it; that doesn't make it fly either. But suppose you take that aluminum, stretch it out in a nice long fuselage with wings, a tail, and a few other parts. **Then it flies, and we call it an airplane**.

Did you ever wonder what makes an airplane fly? Try a few thought experiments. Take the wings off and study them; they don't fly. Take the engines off and study them; they don't fly. Take the little man out of the cockpit and study him; he doesn't fly. Don't dwell on this the next time you're on an airplane, but an airplane is a collection of non-flying parts! Not a single part of it flies!

What does it take to make an airplane fly? The answer is something every scientist can understand and appreciate, something every scientist can work with and use to frame hypotheses and conduct experiments. What it takes to make an airplane fly is creative design and organization. (*What Is Creation Science*, 1987)

CHAPTER THREE

Homology: Evidence for Evolution or Creation?

This is a good time to discuss **homology,** or similarities between organisms. This subject has drawn more people, and some scientists, to form conclusions favoring evolution than practically any other topic. For instance, the limb bones from different mammals do indeed show certain resemblances. Notice the resemblances in the limb bones of some different animals in Figure 12.

More recently the concept of homology has become a very important part of molecular biology. Now scientists are looking at similarities among molecules. Some scientists think molecules should show evolutionary relationships. We will discuss both concepts in this chapter.

Similar by Design or Chance?

Many feel that the relationships shown in the drawing above developed through the ages by evolution from a common ancestor. This idea of "likeness" can be carried beyond the animal kingdom and into the plant kingdom. Notice the drawings comparing the reproductive parts of the sporophyte to leaves and flowers. On the other hand, there seems to be an unanswered challenge to many of the ideas about evolution and homology, especially among serious students of comparative anatomy. A world famous professor of embryology, Sir Gavin de Beer, (F.R.S), said this about homologous

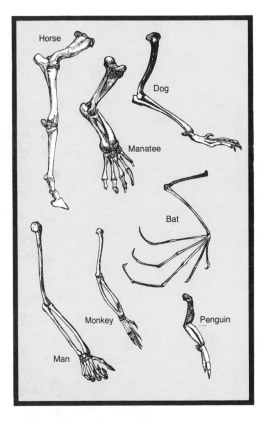

Figure 12 These limb bones from different mammals are said to be homologous, or similar in structure.

23

structures: "*...what mechanism can it be that results in the production of homologous organs, the same patterns, in spite of their not being controlled by the same 'gene'?"* In other words, homologous structures don't come from the same embryological structure; and therefore, they cannot have the same genetic origin. (de Beer, 1971)

For a discussion of how weakly organized the argument is, see Dr. Michael Denton's work titled "The Failure of Homology." Dr. Denton is a research scientist and an M.D. with a Ph.D. in molecular biology, giving him an expertise in comparative anatomy and embryology. He sees little significance in homology. For example, consider the similarity between vertebrate animal fore-limbs and hind-limbs. Dr. Denton writes, *"Yet no evolutionist claims that the hind-limb evolved from the fore-limb, or that hind-limbs and fore-limbs evolved from a common source."* (Denton, 1985)

When we examine similarity in structure, we can't find the kinds of **connections** that are needed to confirm evolution. The creation scientist argues that all life was created to live on the same planet. The structure and function of parts of different organisms are there for a purpose. They work for the benefit of different organisms in environments that are similar. It seems reasonable that there would be different DNA (gene) coding for different traits or similar coding for structures that look alike. There would be no reason for a master designer not to use similar patterns. They show **symmetry** (order), **purpose** and **interdependence**.

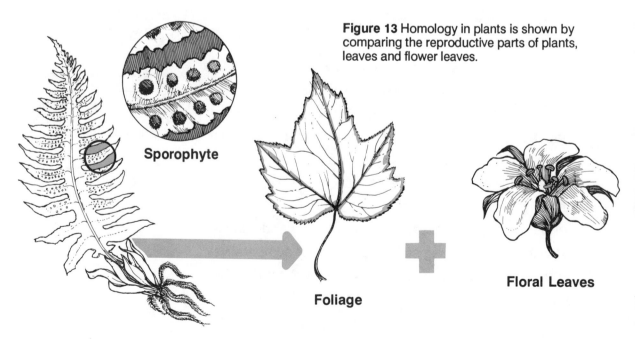

Figure 13 Homology in plants is shown by comparing the reproductive parts of plants, leaves and flower leaves.

Sporophyte

Foliage

Floral Leaves

What About Homology and Molecules?

Looking at homology from a molecular point of view will add interesting fuel to the fire generated by the question of homology. Evolutionists look for sequences in certain molecular structures that will give them an idea of the direction that evolution went. They look at proteins and DNA molecules and **speculate** which ones have been derived from a common ancestral molecule.

The creationist uses the same data to show the uniqueness of organisms to their own kind. For example, you would expect organisms designed for a certain environment to have various degrees of similar biochemistry.

In the 1950's, scientists discovered how proteins were made from amino acids and how the DNA molecule codes the formation of the different proteins. Proteins, such as the blood protein hemoglobin, were found to vary slightly from one species to another and this difference could be measured. It was found that where large structural differences existed between organisms, there were also significant differences in their protein sequences. For example, the hemoglobin amino acid sequences between man and dog differ by 20 percent, while the difference between man and carp (fish) differs by about 50 percent. Finally, evolutionists thought, here was a way to document the evolution of life through molecular biology.

In the 1960's these dreams of scientific evidence for evolution were shattered. It became increasingly obvious that molecules were not going to show an evolutionary relationship, but instead were showing relationships in which the evidence for evolution is absent. As Dr. Denton has observed, "There is not a trace at the molecular level of the traditional evolutionary series: fish to amphibian to reptile to mammal. **Incredibly, man is as close (in hemoglobin differences) to the lamprey as are fish.**" (Emphasis added, Denton, 1985)

Percent sequence difference

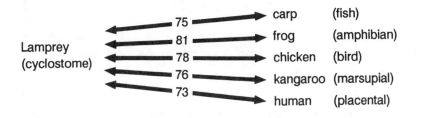

Figure 14 Molecules do not show evolutionary relationships. Hemoglobin shows, according to percent differences, that man is a closer relative of the lamprey than are fish! Molecular homology depends on how one manipulates and arrranges the data.

25

EVOLUTION IN — EVOLUTION OUT

(Input always equals output)

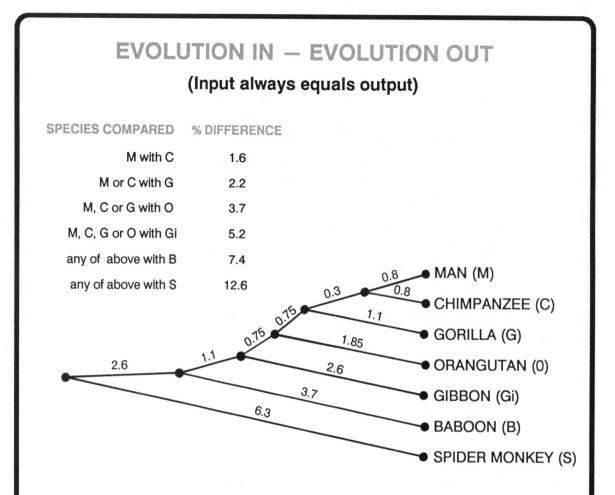

SPECIES COMPARED	% DIFFERENCE
M with C	1.6
M or C with G	2.2
M, C or G with O	3.7
M, C, G or O with Gi	5.2
any of above with B	7.4
any of above with S	12.6

Dr. Vincent Sarich made the following claim about the figure above: "The tree at the right (above) is the only possible arrangement of the above data. If you don't believe this, try to produce an alternative which also fits the data." (Sarich, 1986) Now, what can the creationist say about this apparent relationship? Here is where freedom to explore brings us to a correction in the evolutionist's assumptions. The creationist says that these numbers pointing back to a point of origin are being forced into an evolutionary scenario. If you give the computer some numbers and ask it to sort out numbers back to the origin and build an evolutionary tree, it will always do it. The "program" will create fictitious numbers (or organisms) to fill in the gaps. **We do not have fossils of molecules. The only data we have are based on present-day living organisms.** The tree and the numbers supplied by Sarich are therefore purely hypothetical. As far as science knows, these missing links do not exist.

A creationist model contends that the relationships should center around certain families or orders: man, ape, dog, etc. Only then can you determine how they may be related. This is an alternative view of the same data.

HOMOLOGY—A THEORY IN CRISIS

A bacterium is a single-celled organism that has no nucleus. Yeast is also a single-celled organism, but it has a nucleus. We shall compare the percent difference in amino acid sequences of Cytochrome C in various organisms. First, what is meant by percent difference in amino acid sequences? There are twenty amino acids that are the building blocks of all cells. An average protein might contain several hundred amino acids, but let us use a sequence of only 10 amino acids in our example of two proteins as follows:

Amino Acid Sequences

Protein A	. . . .	Protein B
a	. . . .	a
d	. . . .	d
d	. . . .	d
b	. . . .	**e**
h	. . . .	h
a	. . . .	a
d	. . . .	d
b	. . . .	b
n	. . . .	n
a	. . . .	a

The letters in each column represent different amino acids in two similar proteins. Notice that nine of the ten positions are identical in both columns, so there is a 10 percent difference in the sequences.

The following table shows the percent sequence difference between a bacterium and six other organisms for the protein "Cytochrome C":

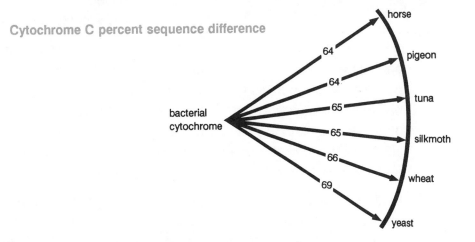

Cytochrome C percent sequence difference

What a surprise for someone expecting Cytochrome C homology to prove evolution! The actual data show the exact opposite. There is less difference between a bacterium and a horse than between a single-celled bacterium and a single-celled yeast. (*Darwin's Enigma: Fossils and Other Problems*, Sunderland, 1987 and *Evolution: A Theory in Crisis*, Denton, 1985)

The Biogenetic Law

Almost every student in biology has heard of "ontogeny recapitulating phylogeny" (the biogenetic law), or the embryo recapitulating (repeating) the evolutionary history of man. Interestingly enough, this whole idea started out as a fabrication by biologist Ernst Haeckel in the late 1800's. In spite of this, the recapitulation theory still appears in many biology texts today. An example of the embarrassment and frustration to biologists lies in the fact that there is little relationship, genetically, as well as in form, between embryos of different organisms. Since there is no conceivable genetic relationship, this leaves evolutionary hopefuls at a loss.

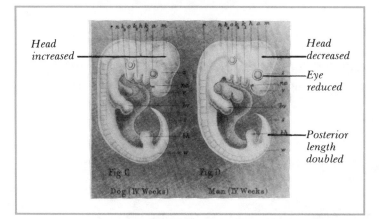

Head increased — *Head decreased* — *Eye reduced* — *Posterior length doubled* — Fig C — Dog (IV Weeks) — Fig D — Man (IV Weeks)

Hackle's drawings were made to show the resemblance of the dog and human embryos, and first appeared in the German edition of the popular National History of Creation in 1868. They were exposed as fraudulent, by Wilhelm His, in 1874.

"Haeckel stated that the ova and embryos of different vertebrate animals and man are, at certain periods of their development, all perfectly alike, indicating their supposed common origin. Haeckel produced well-known illustrations showing embryos at several stages of development. In this he had to play fast and loose with the facts by altering several drawings in order to make them appear more alike and conform to the theory." (*In the minds of Men*, Taylor, 1987)

Evolutionists say "There's no evidence of creation in the human embryo. Otherwise, why would a human being have a yolk sac like a chicken, a tail like a monkey, and gill slits like a fish? An intelligent creator should have known that human beings don't need those things." This belief slowed down scientific research for many years. If you believe something is useless, a nonfunctional leftover of evolution, then you don't bother to find out what it does. The question is, are these *really* yolk sacs, tails, or gill slits? If they are, they would certainly be useless to human beings. Let's take a look at each of these and see just how useless they are.

The Yolk Sac — In a chicken, the yolk sac contains food the chick needs to develop. Humans are nourished in their mother's womb, through the umbilical cord. In a human embryo, the so-called "yolk sac" is the source of the embryo's first blood cells, and death would result without it.

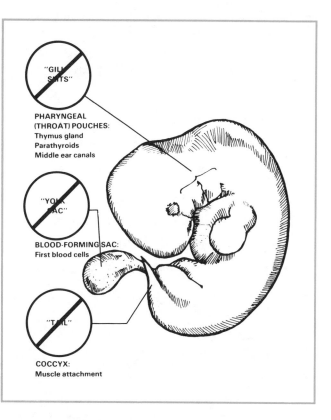

PHARYNGEAL (THROAT) POUCHES:
Thymus gland
Parathyroids
Middle ear canals

BLOOD-FORMING SAC:
First blood cells

COCCYX:
Muscle attachment

Gill Slits — In human embryos at one month, there are pouches in the skin at the neck area. These throat (or pharyngeal) grooves and pouches, falsely called *gill slits*, are not mistakes in human development. While they may appear to resemble the gills in a fish embryo, they have nothing to do with respiratory function. The middle ear canals come from the second pouch, and the parathyroid and thymus glands come from the third and fourth. Without a thymus, we would lose half our immune systems; without the parathyroid, we would be unable to regulate calcium balance and could not even survive.

The Tail — This one has a slight twist to it, since man has a "tail bone" (also called a coccyx), but that is where its similarity ends. The coccyx is one of the most important bones in the body. It's an important point of muscle attachment required for our distinctive, upright posture. In a one-month-old embryo, the end of the spine sticks out noticeably, but that's because muscles and limbs don't develop until stimulated by the spine. As the legs develop, they surround and envelop the coccyx, and it winds up inside the body.

The human embryo is not just human, but also a special unique individual. Eye color, general body size, and perhaps even temperament are already present in DNA, ready to come to visible expression. (*What is Creation Science*, Parker, 1987)

STOP AND THINK

Using descent from a common ancestor to explain similarities is probably the most logical and appealing idea that evolutionists have. Isaac Asimov (1981), a well known science fiction writer, is so pleased with the idea that he says our ability to classify plants and animals on a groups-within-groups hierarchical basis virtually forces scientists to treat evolution as a "fact." In his enthusiasm, Asimov apparently forgot that we can classify kitchen utensils on a groups-within-groups basis, but that hardly forces anyone to believe that knives evolved into spoons, spoons into forks, or saucers into cups and plates.

After all, there's another reason in our common experience why things look alike: it's **creation according to a common design**. That's why Fords and Chevrolets have more in common than Fords and sailboats. They share more design features in common. (*What is Creation Science?* Parker and Morris, 1987)

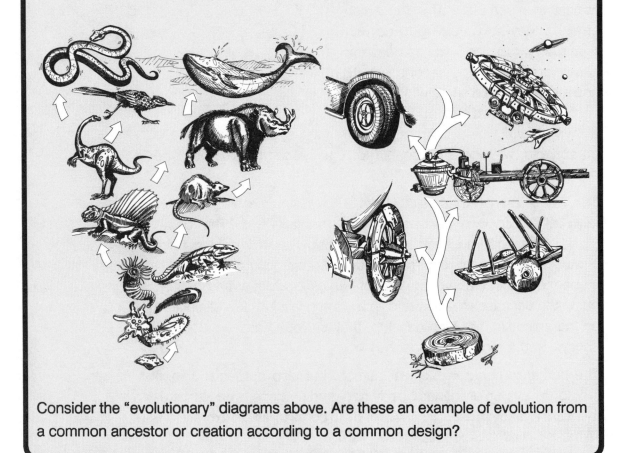

Consider the "evolutionary" diagrams above. Are these an example of evolution from a common ancestor or creation according to a common design?

CHAPTER FOUR

Evolution and Creation Mechanisms

"Mechanisms" for creation and evolution models is a term that is a little far fetched. Actually, scientists have never seen evolution produce a new kind of organism through random processes and creationists have never seen the Master Designer at work putting the organisms together. In spite of this, we can evaluate approches to how the different models may work.

The evolutionary scenario suggests that at the very beginning of life, the ultimate combination of time and chance results in a path from molecules to man. The "mechanisms" that some think are set in motion to produce evolution are: **variation, mutation, migration, isolation,** and **natural selection**. After the first cell arrived these mechanisms helped organisms to adapt and eventually form new species and new kinds of organisms over and over again throughout time.

A Moth Is A Moth

A popular example given by evolutionists is the moth called *Biston betularia*. This moth comes in both dark and light forms. Before 1845, the moth was noted to be mostly light-colored near Birmingham, England. A light colored moth with a light colored background made for good protective camouflage. In the beginning a few of the moths were dark colored (known as the melanic or carbonaria form). When the industrial revolution came along the light trees became sooty and dark. Soon it

Figure 15 "Evolution going on today" . . . that's what many people say about the peppered moth. Because of a change in the color of their background, the light moths so common in 1850 (well-camouflaged in the top photo) lost out in the struggle for life to the more "fit" variety (camouflaged by the dark background in the bottom photo). By 1950, most of the moths were the dark (melanic) variety. Can you accept that as "proof of evolution," or do you wonder if there are boundary conditions that limit the amount of change natural selection can produce?

31

was the light colored moths that were being eaten by the birds and the dark ones survived. Evolutionists considered this a classical example of natural selection leading to evolution of a light to dark *Biston betularia* moth.

The creation model has a different view of these same data. The creationist scenario begins with the basic kind of organism, that is, a dog, a frog, a bird, and so on, being created. From this beginning we can imagine much genetic variation. **Variation is consistent with the creation model** and with the science of genetics. Is the peppered moth really an example of evolution? In reality, *natural selection and evolution are not the same.* At the beginning of our peppered moth story, what did we start with? Dark and light peppered moths. After 100 years of natural selection, all that was changed was the **percentage**, the ratio of dark and white moths. The moths are still moths; in fact, they are still the same species. These moths are an example of how **adaptation can increase the numbers of organisms in a changing environment.**

Stasis — A Creation Model

Laboratory experiments with rapidly multiplying organisms, such as bacteria and fruit flies, show much genetic potential for variation but never a new kind of creature. Everything we see shows **stasis** (an organism that shows little or no change displays stasis.) **Stasis is a creation-model prediction that different biological organisms have uncrossable units.** Creationists point out that natural selection is a conservation mechanism. Natural selection often removes the harmful and injured organism and works to keep the gene pool healthy. Every organism has its own set of genetic traits and these traits have a purpose in the environment.

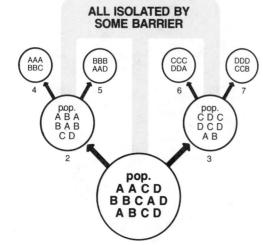

Figure 16 The diagram above shows how original gene pools can be separated. By isolating the different gene characteristics and letting natural selection select out the most favorable (survival of the fittest) new varietes could be formed. Under normal conditions, genetic information can be taken away, but not added to.

Gene pool: the collection of genes that exist in each reproducing kind of organism.

32

Creationists predict that the further scientists explore the more their studies will reveal symmetry, order and interdependence . . . evidence for a Master Designer.

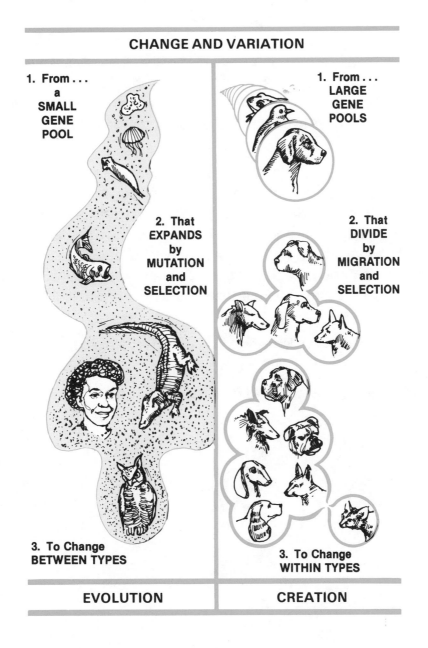

CHANGE AND VARIATION

1. From . . .
a
SMALL GENE POOL

2. That EXPANDS by MUTATION and SELECTION

3. To Change BETWEEN TYPES

EVOLUTION

1. From . . . LARGE GENE POOLS

2. That DIVIDE by MIGRATION and SELECTION

3. To Change WITHIN TYPES

CREATION

Figure 17 Change? Yes—but what kind of change? What's the most logical inference, or the most reasonable extrapolation, from our observations: unlimited change from one type to another (evolution), or limited variation within types (creation)? Given the new knowledge of genetics and ecology, even Darwin, I believe, would be willing to "think about it." (What is Creation Science?, Morris and Parker, 1987)

Red Rocks, White Tails, and Yellow Pines

In northern Arizona, the Grand Canyon acts as a barrier separating two populations of squirrels. The white-tailed, black-bellied population north of the canyon (Kaibab squirrel) is confined to the Kaibab Plateau—an area 60 miles long and 35 miles wide, ranging from 8,000 to about 9,000 feet in elevation. The population south of the Grand Canyon (Abert squirrel—dark tails and white bellies) is found in the higher areas of the Coconino Plateau and southward into central Arizona. Because of differences in color between the two populations separated by the Grand Canyon, evolutionists have used this creature as "a classic example of the role of geographical isolation in evolution."

The Kaibab squirrel is considered to be the most handsome of all squirrels. The flash of its white tail in the green of the ponderosa pines exposes the squirrel as it jumps from limb to limb. Careful examination of the Kaibab squirrel reveals the white tail is streaked with a central line of dark hairs. Except for the nearly pure black belly and the white tail, the Kaibab is indistinguishable from the Abert. Furthermore, in the Kaibab Plateau there are squirrels with a color very similar to the Abert squirrels (they are called "Abert-like-Kaibabs") and in the Coconino Plateau are found some squirrels that have a color similar to the Kaibab squirrels and they are called "Kaibab-like-Aberts"

Is this an example of evolution? Or is the variation due to isolation consistent with the creation model?

Reference: *Grand Canyon: Judgment and Promise*, Institute for Creation Research. (in press)

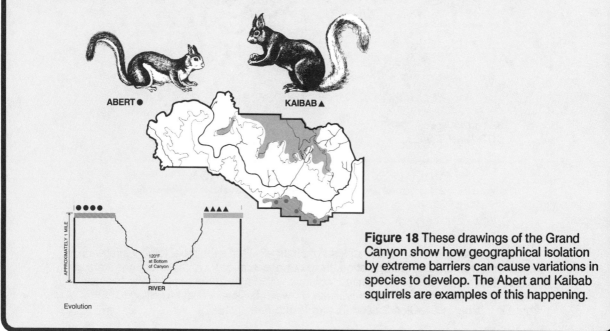

ABERT ●   KAIBAB ▲

APPROXIMATELY 1 MILE

120°F
at Bottom
of Canyon

RIVER

Evolution

Figure 18 These drawings of the Grand Canyon show how geographical isolation by extreme barriers can cause variations in species to develop. The Abert and Kaibab squirrels are examples of this happening.

CHAPTER FIVE

Fossils and the Geologic Column

The geologic layers of rocks and the fossils they hold can be dealt with in this chapter only briefly. Much more detailed volumes are available on these subjects. Let me recommend a few: *Darwin's Enigma: Fossils and Other Problems, Evolution: Challenge of the Fossil Record, Fossils: Key to the Present,* and *The Origin of Major Invertebrate Groups.*

These books are important to your complete understanding of fossils and the geologic column. This chapter, however, will give you a better picture of what is known, and what is not known; you can compare this to what is being taught in the classroom. You may even be surprised to find that many teachers and textbook authors fail to do adequate research on this important issue.

Figure 19 The break in the earth's crust at the Grand Canyon exposes more of the earth's strata than any other place in the world, but less than half the geologic systems are included; there are gaps in the ideal column sequence.

An interesting part of the story about the geologic **layers** is how the term **geologic column** originated. Did you know that there isn't one place in the whole world where you can see this column? But virtually all textbooks in geological science show you this column from the bottom to the top without explanation that it is an **imagined** ordering of the rock strata.

One of the deepest surface exposures to the geologic rock layers is found in the Grand Canyon. This break in the earth's crust exposes more layers than any other place in the world; yet, **less than half of the geologic systems are included!** There are major gaps in the "ideal" or imagined geologic column sequence that is pictured in textbooks. If we are to understand the geologic column better, we must keep this fact of science in mind.

Index Fossils and Circular Reasoning

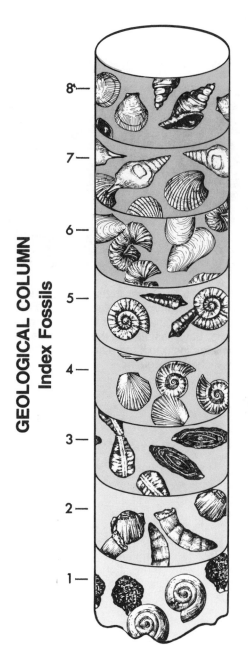

There are twelve major layers that form what is called the **standard geologic column**. Each of these layers is identified by the fossils that are found in it. But, strangely enough, most often the fossils are dated by the strata in which they are found. Can you see the faulty logic in this approach? Let us read what some evolutionary geologists say. Dr. J. E. O'Rourke, writes about this in "Pragmatism Versus Materialism in Stratigraphy," for the **American Journal of Science**. (1976)

> The rocks do date the fossils, but the fossils date the rocks more accurately. Stratigraphy cannot avoid this kind of reasoning . . . because circularity is inherent in the derivation of time scales.

This is amazing! Fossils are used to date the rocks and the rocks date the fossils. What can be the value of such a chronology? Professor of geology, D. Ager, writes about his disappointment when physicists say that strata are dated radiometrically. He said that he could think of no cases where radioactive decay methods are used to date fossils. (Ager, 1983)

This dating process for layers and fossils is called the **index fossil system** (see Figure 20). For example, if a rock layer contains mostly fossils of sea animals called *trilobites* and *lampshells*, then these layers are part of the Cambrian (kame-bree-an) system. Using this system, the different layers of the strata can be given a label and with this label comes a length of time that the index fossil is assumed to have evolved.

Figure 20 Index fossils: certain fossils that identify strata or earth layers.

GEOLOGICAL COLUMN
Index Fossils

What Do Fossils Really Tell Us?

The **Cambrian** and the **Pre-Cambrian** fossils have always brought on a storm of criticism. Axelrod (1958), a paleontologist, said:

> One of the major unsolved problems of geology and evolution is the occurrence of diversified, multi-celled marine invertebrates in the lower Cambrian rocks on all the continents and their absence in rocks of greater age.

Addressing the absence of fossils in rocks of greater age, he went on to say:

> However when we turn to examine the Pre-Cambrian rocks for forerunners of these early Cambrian fossils, they are nowhere to be found. Many thick (over 5,000 feet) sections of sedimentary rock are now known to lie in unbroken succession below strata containing the earliest Cambrian fossils. These sediments apparently were suitable for the preservation of fossils because they are often identical with overlying rocks which are fossiliferous, yet no fossils are found in them.

This evolutionary paleontologist was willing to ask some hard questions about **transitional forms** showing how one organism turned into another. Ager, an evolutionary geologist who was quoted earlier, seems to think this is a problem all the way through the fossil record. He predicted that no matter where we searched we would find *"not gradual evolution, but the sudden explosion of one group at the expense of another."* This is what the creationist model predicts. Creationists are saying that if instantaneous creation took place then we should see evidence of complex, fully developed organisms in the geologic layers as a record of this explosion of life. Many scientists say that a "falsifying test" for creation and evolution lies in the fossils themselves. The fossil record as well as living organisms show no transitional forms. Does this falsify the evolutionist or the creationist claims?

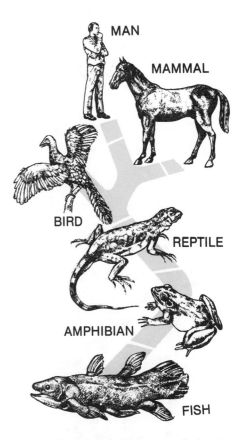

Figure 21 An "evolutionary tree," showing the alleged relationship among vertebrate groups.

Figure 22 *National Geographic* writes of the transition from hairy, four-legged mammals to whales. Whales are mammals with some fish-like traits (see Dec. 1976, *National Geographic* insert)

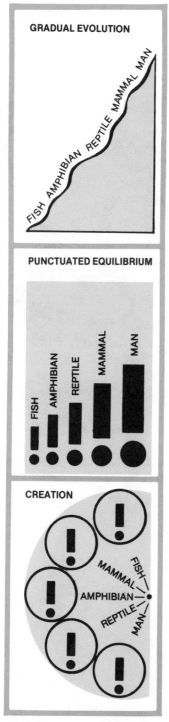

Figure 23 Followers of the punctuated equilibrium idea agree with creationists that fossils show only limited variation within separate kinds. They attempt to explain these observations by offering a new evolutionary model.

The Punctuated Equilibrium Theory

Knowing about significant gaps in the fossil record, evolutionists have come up with an explanation called "punctuated equilibrium".

Professors Stephen Gould and Niles Eldredge (1980) introduced a hypothesis which they feel could explain gaps in the fossil record. Certain species appear in the fossil record fully formed and persist for a long time; they then disappear from the record looking very much as they did in the beginning. Later, other species appear in the fossil record fully formed. The change to a different species occurred suddenly. These fossils, they say, are the offspring of the ones that had disappeared. In other words, evolution had been going on all the time, but the evidence can't be observed. (Stanley, 1979; 1981)

Assumptions are not facts of science. The gaps are there and there is no scientific principle that will help to bridge them. Creationists point out that everything we know about the genetic mechanisms excludes this idea from rational scientific thought.

Jellyfish and Fish

What is the evidence from the fossils for the origin of invertebrates (animals without backbones) and vertebrates (animals with backbones)? Evolutionists say that vertebrates evolved over 100 million years from invertebrates. This, of course, is a major jump, outside known biological processes, and one that brings strong counter arguments from the creationists. Could an organism such as a jelly fish evolve into a bony fish? The evolutionary hypothesis for this idea is somewhat complex. Evolutionists claim that the transition from invertebrate to vertebrate passed through a simple chordate (animals with a soft spine or notochord) state.

> How this earliest chordate stock evolved, what stages of development it went through to eventually give rise to truly fish-like creatures, we do not know. Between the Cambrian, when it probably originated, and the Ordovician, when the first fossils of animals with really fish-like characteristics appeared, there is a gap of perhaps 100 million years which we will probably never be able to fill. (Ommaney, 1964)

It seems that even the well-known advocate and teacher of evolution, Dr. A. S. Romer, believes that all of the major fish classes are clearly and distinctly set apart from one another. There are no transitional forms. He said:

> In sediments of the late Silurian and early Devonian age, numerous fish-like vertebrates of varied types are present, and it is obvious that a long evolutionary history had taken place before that time. But of that history we are mainly ignorant. (Romer, 1966)

The conclusion—Not one single transitional form between invertebrates and fish has ever been found!

A Fish Story

According to the evolutionary scenario the **fish** gave rise to the **amphibia** (frogs, salamanders). As Carl Sagan suggested in his Cosmos television series, during a drought in the Devonion, when the swamps were drying up, some fish would have found it very convenient to have evolved feet and legs for walking overland (Sagan, 1980). The idea is that a lobe-finned fish (crossopterygian) was supposed to have evolved into an amphibian (ichthyostegid) about 500 million years ago. Do we have any evidence of fossils that indicate such a transition between fish and amphibians? No! Not one fossil has ever been found showing part fins

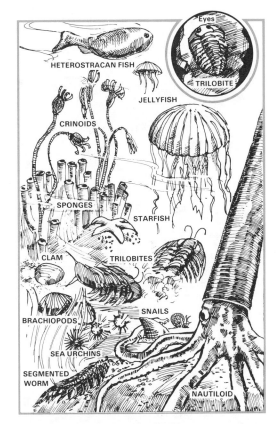

Figure 24 The simplest community of abundant fossils, the "Trilobite Seas" (Cambrian System), contains almost all the major groups of sea life, including the most complex invertebrates, the nautiloids, and the highly complex trilobites themselves (insert above). Darwin called the fossil evidence "perhaps the most obvious and serious objection to the theory" of evolution.

39

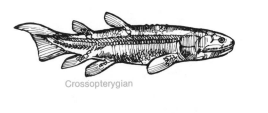

Crossopterygian

Ichthyostega

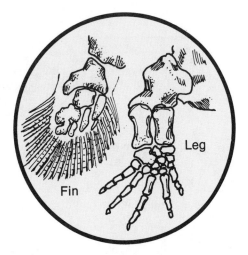

Leg

Fin

Figure 25 Note similarities and differences in the bone pattern (circled, upper left) for the fin of a crossopterygian fish and the leg of a fossil amphibian, Ichthyostega.

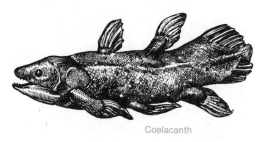

Coelacanth

Figure 26 The coelacanth, a crossopterygian, has been found alive off the coast of Madagascar.

and part feet. The ichthyostegid amphibian had your basic amphibian limbs. When Dr. Colin Pattersen, an Associate at the British Museum of Natural History, and author of the text **Evolution**, was asked whether he thought the crossopterygian was the ancestor of the ichthyostegid, he answered, "I have questions about that . . . It is futile to be looking for answers to questions which we have no way of answering." (Sunderland, 1988)

Again there is no fossil evidence. The following are just a few of the serious flaws in the fish-to-amphibian theory:

1. The bone pattern in the lobe-finned fish appears abruptly and complete in the fossil evidence. No fossils connect this bone pattern to the fins of other fish.

2. There is no elbow joint in the fin. No fossils of fish with jointed fins have been found.

3. The pelvic (hip) bone of the fish is small. This bone is loosely attached to muscle and does not connect to the back bone.

4. Other orders of amphibians appear in the same geologic layers.

5. The "living fossil" lobe-finned fish, **coelacanth**, is adapted for life in the deep sea, and does not use its fins for walking.

As we have seen many times before, the data suggests abrupt, sudden appearance of fully developed, complex organisms. **Only imagination in the realm of science fiction can create evolutionary scenarios from the fossil data.**

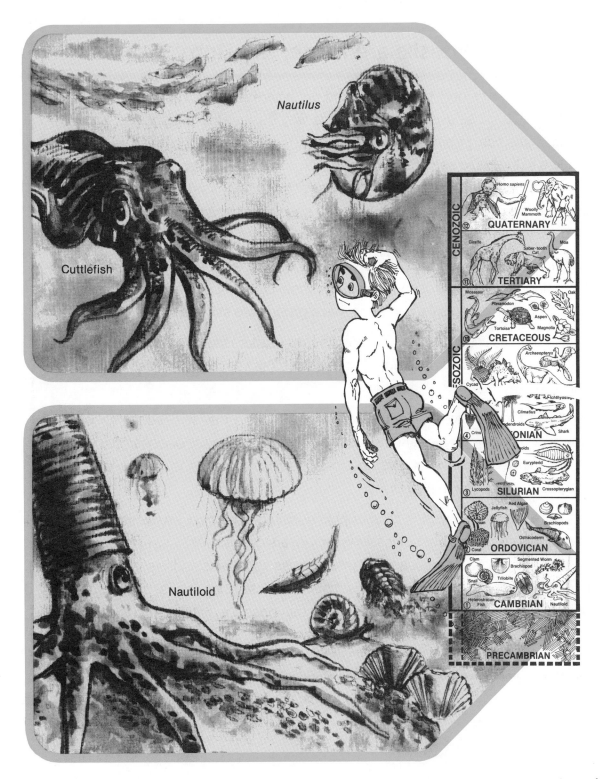

Figure 27 Many complex invertebrates are found in Cambrian strata; the ones living today are similar. What does the comparison mean to you?

41

Fossils are Made, Not Born

Suppose you had a burning desire to find out where snails came from. You search the fossil evidence all over the world, all the way back to the "beginning," and sure enough, snails come from snails. Where did the most complex of all the invertebrates, members of the squid and octopus group, come from? Again, you search through all the fossil evidence, all the way back to the very "beginning," and sure enough, squids come from squids. In fact, the "first" squids, the nautiloids, are a bit more impressive than most modern forms. And, of course, trilobites seem to come only from trilobites.

In other words, you find snails and squids and trilobites as fossils; you don't find "snids" and "squails" and "squailobites," or some other in-between form or common ancestor. The "missing links" between these groups are still missing.

The lack of transitional forms in the fossil record is addressed by the creation model and the evolution model of origins. The evidence best fits which model: Creation or Evolution?

How do you make a fossil? Try this experiment. Take a freshly dead animal off the roadway. Put it in your yard and take notes daily. Soon you will discover two things: first, this animal carcass is not going to produce a fossil and second, you are not going to be popular with your neighbors.

Fossils are formed by sudden catastrophic burial and chemical processes. Further, there are billions upon billions of fossils in rock layers over the earth, layers of sedimentary rock that stretch for thousands of miles across continents and in some instances plunge many miles into the earth's crust. The data suggests a catastrophe of worldwide signifience. What are some of the possible causes for these fossil graveyards?

CHAPTER SIX

Fossils: What Do They Record?

One idea about how the universe and life first come upon the scene is that all things were created by a Creator. The Biblical account of creation as accepted by many Jews, Moslems, and Christians is one of several such accounts which agree with this theory. Many scientists claim that the variety of life we now see could have come from the first original created kinds of organisms. The following pages will show how creationists believe scientific data seem to support these ideas.

The creationist maintains that the record shows that life exploded into existence all of a sudden. He also observes that this record doesn't follow a pattern of simple to complex life, A look at the **strata** reveals that even in the **Cambrian** layers (see Figure 28) many forms of complex life were present, yet ancestors cannot be found in older rocks.

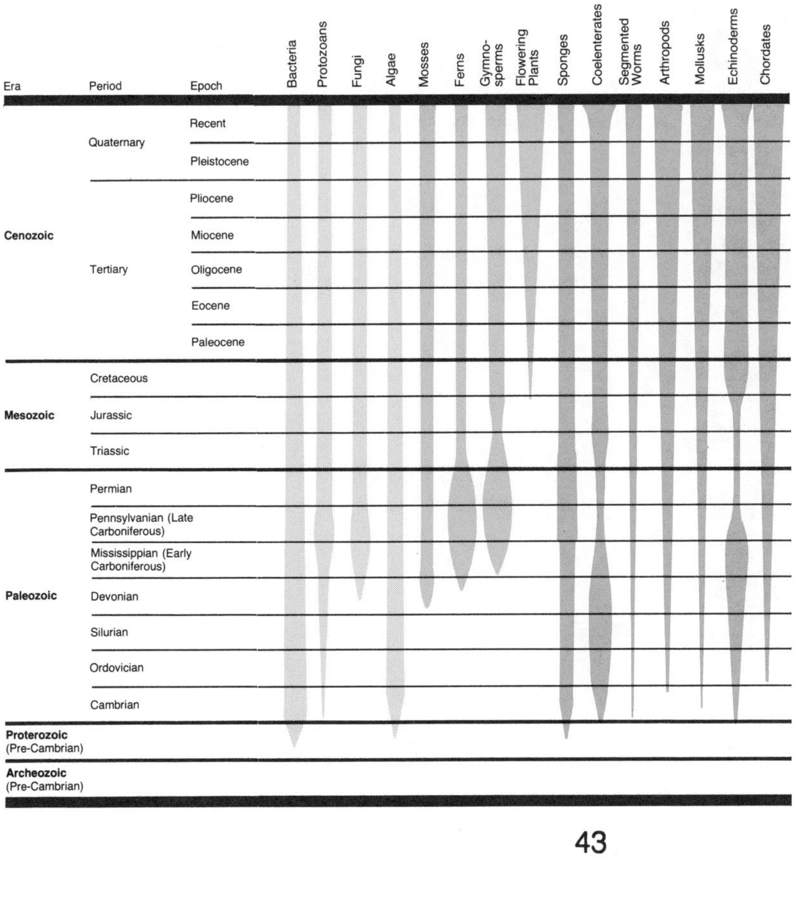

FOSSIL RECORD &
EVOLUTIONARY TIME SCALE

Figure 28 Table shows the fossil record as it is assumed to be found in the geologic column. The sequences shown are evolutionary assumptions.

Many geologists are becoming convinced that most of the vast sedimentary layers of rock covering the earth today were deposited by floods or other major catastrophes. Flood waters came quickly and destroyed almost all of the animal life and much of the land plant life. The evidence for a *global* flood includes continent-wide **sedimentary** formations laid down as a mixture of sand and water along with vast fossil graveyards. This, of course, is in conflict with a gradual formation of the geologic column used as a time table. Many studies made by geologists give good reasons for doubting the standard interpretation of the **geologic record**. These studies indicate that the earth's sediments do not seem to have been deposited over long periods of time, but rather short periods of time. The whole topic of dating and the age of things will be discussed more completely in the chapter on *Time and Earth History*.

A Worldwide Flood

Flood geologists have formed some ideas on how these deposits were made; they state that if a person looks at the record, he will notice that a flood interpretation fits the scientific data very well.

Figure 29 Flood geologists think that this idea is in keeping with the fossil evidence that has been found in the strata.

FLOOD STAGES

STAGE 1
During the violent flood stage the earth's surface was broken up by the bursting of underground water reservoirs, combined with the action of huge waves.

STAGE 2
Torrential rain and the gushing of subterranean water continued until the flood waters covered the earth's highest mountains.

STAGE 3
As the flood begins to quiet down, higher density animals and plants settle out first (sea shells, heavier sediment, etc.).

STAGE 4
Layers of sediment are deposited all over the earth. Oceans assume their present basins.

As the flood waters were rushing forth, it produced a complicated mix of materials, the higher **density** objects, those that weighed more in water, and the objects that were more streamlined, such as sea shells, etc., settled out first. The land animals that could move quickly went to the higher ground. Many of these animals at higher levels were eventually trapped and buried. Fishes, worms, etc., were quickly buried in the sediment (see figure 29). The flood geologist points out that the type of fossils found in a particular area is in keeping with the **ecological zones** (the particular part of a zone community or ecosystem that an organism inhabits). These would be buried first during a flood. The Nebraska fossil graveyards at Agate Springs look something like the picture in Figure 31. Consider the additional facts listed below:

1. Almost all sediments that contain fossils were probably water deposited.

2. Great underwater canyons around every continent in the world indicate a rushing and gouging of water.

3. 750,000 square miles of sediment thousands of feet thick are found on the Tibetan Plateau, three miles high.

4. Consistently wide sedimentary layers of sandstone, limestone, and shale could only have been deposited by water on a large scale.

5. Fossil deposits all over the world can be found out of their assumed evolutionary geologic order. Missouri, Kentucky, and other places are examples of this.

6. The existence of vast fossil graveyards. It has been estimated that the Karoo formation in South Africa contains the fossils of 800 billion vertebrates, including reptiles.

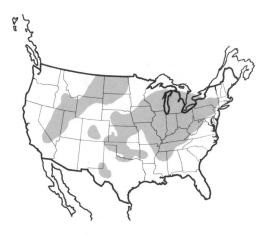

Figure 30 Map shows the wide distribution of sandstone (St. Peter's sandstone) in the United States. This is typical of many water deposits of sediments throughout the world that seem to support a world-wide flood concept.

Figure 31 This rock slab was taken from the well-known "bone bed" at Agate Springs, Nebraska, a stratum in which fossil bones of thousands of mammals have been found. The bone layer runs horizontally for a large distance in the limestone hill, and has evidently been water-laid. Fossils of the rhinoceros, camel, giant bear, and numerous other exotic animals are found jumbled together in this stratum.

World's most widespread coal seam (Desmonesian Lower Kittanning, Colchester, Bevier coal)

Area covered by coal equals approx. 160,000 square miles.

Recent experiments have shown that coal and oil can be formed in a very short time (minutes to hours) under heat and pressure. Coal formations were no doubt formed from vast amounts of plant material transported by water and sediment. Finally, the material was buried. Perhaps this tells us something about the age of our coal fields? (After Dr. Steve Austin).

CHAPTER SEVEN

Missing Links or Just Missing?

Archaeopteryx — Fowl or Foul?

As lecturer and author Ken Ham is fond of saying, "The only thing for certain about missing links is that they are missing." If there really are no transitional forms in the fossil record, then how do creationists explain the famous **Archaeopteryx**, the fossil that supposedly shows that birds evolved from reptiles. Among the specimens of Archaeopteryx that have been found, there is only one very complete specimen. The fossil imprint (from the Jurassic Solnhafen limestone in Germany) is shown in the drawing below. (Recently, this fossil has been questioned as possibly being a fraud).

Notice the bird-like features: feathers, wings and bill. Notice also the reptile-like features: a long bony tail; claws on the wings, and socketed teeth in the bill. The lack of a breastbone and backbones that are not fused are also more reptile-like than bird-like. Evolutionists claim that these features are exactly what one should expect in a transitional form.

Once again, the creationists claim, the evolutionist has overstepped the limits of the data available. The feathers that have been associated with Archaeopteryx are of several kinds. Experts claim that the primary feathers show, without question, that this was a strong flying bird. (Olsen, 1965) There is nothing about Archaeopteryx that would prevent their flying. (Fuduccia and Tondoff, 1979). Furthermore, feathers are quite

Figure 32 The famous Berlin fossil specimen of Archaeopteryx (a), and an artist's conception (b).

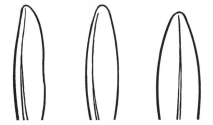

Figure 33 Flight feathers of Archaeopteryx (center) resemble those of strong fliers (left) rather than those of non-fliers (right).

47

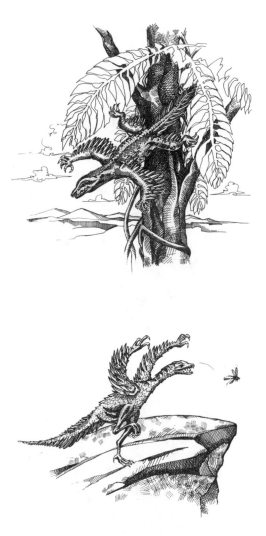

complex structures. They have hooks and eyelets for the zippering and unzippering necessary for efficient flight. The Archaeopteryx feathers offer no clues about how a reptile scale could evolve into a bird feather.

But what about teeth, wing claws and the lack of a breastbone? Again, a closer study of the data shows that these are not really as reptile-like as first supposed. The ostrich, for example, has claws on its wings that are even more "reptile-like" than those of Archaeopteryx. The bird, *hoatzin*, does not have much of a breastbone and although no living birds have socketed teeth, there are examples of extinct birds that did. (Parker and Morris, 1987)

Perhaps the final argument against Archaeopteryx as a transitional form has come from a rock quarry in Texas. Here scientists from Texas Tech University found bird bones encased in rock layers farther down the geologic column than Archaeopteryx fossils. (Beardsby, 1986)

Archaeopteryx cannot have been the ancestor of birds because birds already existed!

A Horse Story

Another example of fossils used to show evolutionary transition is the famous horse series. Once again, much like the "geologic column," many museums and most textbooks describe the evolution of the horse from beginning to end just as though the fossils exist that way somewhere. Well, the fact is that this series is nowhere to be found as you see it.

Textbook writers have been told for many years that the horse series is unquestionable; therefore, it is automatically included in texts as a fact of science. Dr. Garrett

Figure 34 Two concepts of the evolutionary ancestor of birds, called "pro-avis," are redrawn here from an article by Yale's John Ostrom (1979). As Ostrom says, "No fossil evidence of any pro-avis exists. It is a purely hypothetical pre-bird, but one which must have existed." But this case for evolution is based on faith, not facts. The fossils found so far simply show that birds have always been birds, of many distinctive types.

Hardin, writes in "Nature and Man's Fate," 1960, pp. 225-6:

> There was a time when the existing fossils of the horses seemed to indicate a straight-lined evolution from small to large, from dog-like to horse-like, from animals with simple grinding teeth to animals with complicated cusps of the modern horse . . . As more fossils were uncovered, the chain splayed out into the usual phylogenetic net, and it was all too apparent that evolution had not been in a straight line at all. Unfortunately, before the picture was completely clear, an exhibit of horses as an example . . . had been set up at the American Museum of Natural History, photographed, and much reproduced in elementary textbooks (where it is still being reproduced today).

A clear understanding of these fossils can be achieved when you discard the imagined ideal, which doesn't appear anywhere. It is important to know that fossils of **Eohippus** (dawn horse) have been found in surface strata alongside two modern horses, **Equus nevadensis** and **Equus accidentalis.**

Some museums are excluding **Eohippus** in the horse series entirely. When the fossil was first found it was classified **Hyracotherium** because the skeleton was said to be identical to the rabbit-like hyrax living in Africa today. (Sunderland, 1988)

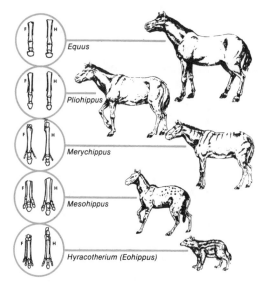

Figure 35 The series above shows a reduction in the number of hoofed toes, an increase in size, and lengthening of the face. Genetically, we start with a horse and end up with a horse.

Figure 36 Both horses here are mature adults! Can you think of other examples of great variation within a group of plants or animals?

STOP AND THINK

At last—evidence of evolution!...*or is it?* The famous Archaeopteryx combines features most often found in reptiles (teeth, claws, unfused vertebrae, and a long bony tail) with features distinctive of birds (wings, feathers, and a furcula or wishbone). Does Archaeopteryx provide clues as to how scales evolved into feathers, or legs into wings? Is Archaeopteryx most likely an evolutionary link, or a mosaic of complete traits (a distinctive created type)?

A **melange** is an animal which has traits that are similar to other animals, but not related to those other animals. The Australian platypus is a living example of an animal, like the fossil Archaeopteryx, that has a mosaic of traits that are seen in other animals: it has a bill like a duck (but the platypus did not evolve from a duck!), it lays eggs like a turtle (but it isn't evolving into a reptile!), it has a tail like a beaver (but it isn't related to the beaver either!), and it uses sound and echo-location to find food like a bat (but is isn't going to ever fly!). The little platypus is a distinctive, created animal adapted to a successful life "down-under." Could Archaeopteryx also be a melange?

50

CHAPTER EIGHT

Ape to Man: Fact or Fiction?

Man's origin has always been the most intriguing question for those studying fossil evidences. How did man come on the scene? From what kind of ape did he evolve? Was man the abrupt creative act of an intelligent designer?

These are legitimate questions and worthy of exploration. Unfortunately, the information from most textbook sources is biased toward evolution only. Our objective in this chapter will be, once again, to bring the significant parts of the evolution and creation models into focus.

"The Missing Links"

Once some scientists believed that they evolved from ape ancestors, they began searching for fossil evidence. Unscientific approaches to this question have resulted in virtually every fossil ape discovered being touted as a discovered "missing link," an evolutionary ape ancestor of man. This approach has also resulted in frauds and faked data being accepted as "missing links." Nebraska man was reconstructed from what was later found to be the tooth of an extinct pig. This evidence was used by Clarence Darrow at the famous Scopes Trial to try to force the teaching of evolution in public schools in Tennessee in 1925. Another fraud, Piltdown Man, was constructed from faked fossils placed in a gravel pit. Both Nebraska Man and Piltdown Man were widely hailed as "missing links" by the scientific community. (Gish, 1986)

If a scientist believes that men came from apes, he must imagine these "apemen" into existence. As a science

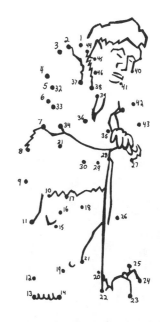

Figure 37 Artist conception of the "Missing Link" based on the fossil evidence.

Nebraska Man

Figure 38 The famous Nebraska Man was drawn from an extinct pig's tooth.

writer recently admitted, "Bones say nothing about the fleshy parts of the nose, lips or ears. Artists must create something between an ape and a human being; the older the specimen is said to be, the more apelike they make it." (Rensberger, 1981)

The Fossil Apes

According to the scientific creation model men and apes were created as separate kinds. Biological variations occur within these created kinds, but nowhere does there exist evidence that apes evolved into humans. The creation model predicts that all ape fossils would be either fossils of living (extant) apes or fossils of extinct apes, which for some reason could not survive. What does the fossil evidence show?

Ramapithecines

Fossils classified as ramapithecines are of two types: **Ramapithecus** (*Rama* is named for a Hindu god and *Pithecus* means ape) and a larger ape, **Sivapithecus** (*Siva* also being the name of a Hindu god). Until recently, only a few teeth and small fragments of fossil skulls had been found. Based on this limited fossil evidence scientists believed they could recognize human-like characteristics in the teeth and skull fragments. In 1979 the crushed skull of a **Sivapithecus** was discovered in the foothills of the Himalayan mountains. The new evidence resulting from the reassembled fragments suggests that the ramapithecines are very similar or identical to the modern orang-utan apes. (Leaky, 1982; Bleibtreu, 1985)

Australopithecines

The extinct australopithecine apes are believed to have evolved from the ramapithecines. Now that the ramapithecines appear to be fossil orang-utans, no evolutionary ancestor for australopithecines has been found. Some scientists believe that **Australopithecus**,

Figure 39 The fossil ramapithecines were once thought to be an ancestor of human beings. Scientists now believe them to be extinct relatives of the modern orang-utan.

52

which means "southern ape," consists of a single species. They figure that the range of fossil skeletal types found can be due to normal individual differences and to differences in the sexes—a larger skeleton for the male and a smaller one for the female. (Johanson and Edey, 1981)

The opposite view is that there were many australopithecine species. Where fossil skeletons of different sizes are found together it is suggested that the different species lived together in the same habitat. The generally accepted names following the many-species hypothesis are *Australopithecus afarensis, Australopithecus robustus (boisei)* and *Australopithecus africanus.* (Leakey, 1981; Mehlert, 1980)

What happened to the australopithecines? Suddenly they appear as fossils in the sediments of Africa and just as suddenly they disappear. Did they simply become extinct like the dinosaur? Did they evolve and change into humans? Or were the australopithecines ancient apes whose surviving descendants can be found among the modern apes?

The creation model affirms that apes have always been apes. Australopithecines were not gorillas or chimps. They were apes, but **not** the same as any modern apes. The well known fossil hunter Richard Leakey has observed that the size of the australopithecine brain was not very different from that of a chimpanzee or gorilla. (Gribbin and Cherfas, 1981) The wear pattern on the teeth enamel of *Australopithecines* is that of a fruit-eater like modern chimpanzees. (Leakey, 1981) The foot bones of *Australopithecus afarensis* are slightly curved, i.e., a bone structure expected in a tree dwelling ape. (Leakey, 1981) Another famous fossil ape hunter, Donald Johanson, has recently observed that the skull of *Australopithecus afarensis* looks like a small female gorilla.

Figure 40 The alleged ability of some fossil australopithecines to walk upright can be observed today in the pygmy chimpanzee

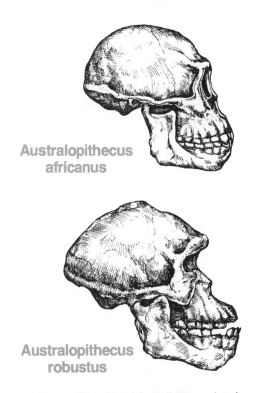

Australopithecus africanus

Australopithecus robustus

Figure 41 Considerable variation exists in fossil Australopithecine skulls. Are they related to modern chimpanzees or gorillas?

An Ape Named Lucy

A research team headed by D. C. Johanson from 1972 to 1977 surveyed the Afar area of Ethiopia. Their research brought to light a skeleton they called Lucy. This skeleton was 40% complete. While Lucy has received much attention, what conclusion can we draw from her skeleton? According to evolutionist and anthropologist Richard Leakey, Lucy's skull was so incomplete that most of it was "imagination, made of plaster of paris."

Well, what is known for sure about Lucy? The teeth have the characteristics of ape teeth and the limited number of skull remains show ape-like cranial features. The only three bones associated with Lucy which are claimed to show any distinctly human features are (1) a knee joint found in sediment **80 meters below** Lucy, (2) the ratio of the arm length to the leg length in Lucy and, (3) Lucy's left pelvic bone.

Lucy's Arm/Leg Ratio

An ape's arm is about as long as its leg. A human's arm is about 3/4 as long. Lucy's arm is listed as 83.9% as long as her leg which, if you think about it, is a rather precise measurement. Lucy, at 83.9%, is somewhere between ape and man. But in making this measurement, a bias was introduced. The leg bone had been broken in two places and one end had some crushing. Johanson, by his own admission, estimated the leg length anyway, which essentially makes this "precise" proof ratio useless.

Lucy's Pelvis

Lucy's left pelvic bone is complete, but according to Johanson it is distorted. Since this bone is the only one of its kind in existence, it is puzzling as to how the experts actually know it is distorted, unless of course they wish to reshape it until it appears able to walk upright. Other scientists, such as Lord Zuckerman and Charles Oxnard, do not believe that Lucy walked upright.

Conclusion

What is Lucy then? There is no evidence that she is an ancestral relative of man. She may probably even be a relative of today's pygmy chimp, which, when not swinging in the trees, frequently walks upright and has striking similarities to Lucy's structure. Still, it is also possible that australoeithecines, like Lucy, are a unique kind of ape-like creature, genetically unrelated to either the present day apes or man, but are now extinct.

Reference: *Little Known Facts About Dead Apes*, Unfred and MacKay, *Creation* (Vol. 8, 1986)

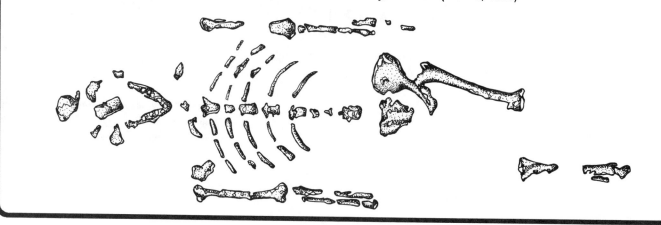

Homo Habilis

In 1946 a fossil australopithecine was discovered which appeared to be different from other australopithecines. It was excavated from a prehistoric butchering site. The site was littered with stone tools and animal fossils such as pig, horse, catfish and tortoise. The australopithecine fossil was scattered: foot bones were among bones of a fossil horse and hand bones were among bones of a fossil pig. Again the evolution model insists that this butchering site must have been occupied before man evolved. Since evolutionists could not accept that the stone tools were made by the horse or the pig, the fossil ape was elected as the tool maker. This ape was named *Homo habilis* which means "handy man." However, more recent re-examination of the finger bones of this fossil ape has led scientists to conclude that the *Homo Habilis* hand was "similar in overall configuration to chimpanzees and female gorillas." (Susman and Stern, 1982)

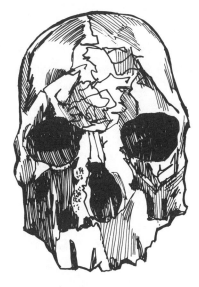

Figure 42 Skull 1470. See section "Dating the World's 'Oldest Man'" for more information about this fossil skull.

Man Tracks

Some scientists believe that the australopithecines walked upright like humans, rather than like chimpanzees, for example. Richard Leakey recently admitted that evolutionists do not know whether or not Australopithecus walked upright because no one has yet discovered a complete skeleton associated with an Australopithecus skull. It is necessary to know exactly how the spine is attached at the base of the skull for an accurate interpretation of upright walking. That evidence is inconclusive. (Cherfas, 1982)

The most convincing evidence of upright walking comes not from fossil skeletons, but from footprints. While the Great Rift Valley was opening along the eastern length of Africa, volcanic activity was increasing. A layer of ash covered the surrounding landscape with each episode of volcanic explosions. When an animal walked across the ash, it left its track. Rain cemented the ash and new ashfalls protected the tracks from further erosion. Human-like tracks occurred along with the numerous animal tracks. The footprints were of an adult and a child. The site is Laetoli in northern Tanzania. (Reader, 1981)

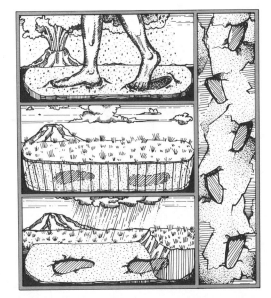

Figure 43 The simplest explanation for "human- like" footprints in the Laetoli ash sediments is that they were made by humans.

According to the evolution model these tracks were made before man had evolved. Therefore, the tracks must have been made by an ape walking upright like humans. Human fossils are found in this region, but evolutionists believe these humans must have lived much later.

A creation model would not reject the hypothesis that the Laetoli tracks could have been made by humans. Man existed during the time of the australopithecine apes. Fossil human bones and stone tools have been discovered along the Great Rift Valley and it is possible that these "human-like" tracks were actually made by humans. (Parker and Morris, 1987)

Human Fossils

The early evolutionists, not having any fossils showing a transition between ape and humans, still imagined what the "missing links" must have looked like. When human fossil skulls were discovered which showed extended brow ridges or brain size, artists gave them apelike features and mannerisms. Human fossils labeled *Homo erectus,* which means "erect man," were initially thought to be a missing link. However the early *Homo erectus* fossils were known only from missing mystery fossils (Peking Man) and hoaxes (Java Man).

Homo Fossils

Some African and Asian human fossils are the remains of individuals whose brain size ranged from 775 to 1300 cubic centimeters (cc). This range is smaller than the average for modern humans, which is typically reported as 1450 to 1500 cc. However, the smaller brain sizes of some *Homo* fossils are within the range of brain sizes known for modern human populations. This range is approximately 830cc to 2300cc. (Gish, 1986) An apparently normal woman with a brain size of 720cc has been documented (Mehlert, 1979). Also, caution should be used when associating brain size with intelligence. It is popular to believe that a larger brain size corresponds with greater intelligence. This assumption is incorrect. Here are several examples why brain capacity is not a satisfactory measure of intelligence:

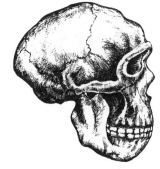

Neanderthal Man

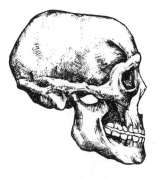

Cro-Magnon Man

Figure 44 Comparison of fossil Neanderthal skull to that of fossil skull more like that of people living today.

1. Individual human brain capacities widely vary (e.g., ranges from 830cc to 2800cc in modern peoples) without a corresponding variation in intelligence;

2. Men average a larger brain capacity than women without a corresponding advantage in intelligence; and

3. Whales, dolphins and elephants all have larger brains than humans.

Some *Homo* fossils were small in stature averaging around 153 centimeters (5-feet) in height. There are people living whose average height is less than the height of these ancient peoples. The Negritos of Oceania have an average height of 147 centimeters and the Pygmies of highland New Guinea are 150 to 157 centimeters in height (Bellwood, 1979). Both the variation in height and variation in brain sizes observed in fossil humans can still be found in today's population. A recent *Homo* fossil discovery reported by Richard Leakey is that of a male youth (estimated to be twelve years old) who, when fully developed, would have been about six feet tall! (*Understanding Genesis*, 1987)

Neanderthal

The picture of dull-witted, shambling, frowning Neanderthal was popular until recently. Now, as an evolutionist has written, "Most paleoanthropologists and the artists working under their direction have given the Neanderthals a shower and a shave and straightened up their shoulder. Neanderthal men and women no longer shuffle along on bent legs, staring vacantly. Now they stride erect and with purpose—not exactly like us in the face, but clearly a race of our own kind." (Rensberger, 1981)

The evolution model holds that increasing brain size is a characteristic of human evolution. The larger the fossil brain size is calculated to be, the closer it is assumed to be to modern humans. (Lewin, 1982) It is often taught that a characteristic of future man will be a larger brain size than humans of today. This story is popular, but the scientific evidence is against it. The average brain size for modern humans is around 1500cc, but the average brain size calculated for Neanderthal-type people is larger—around 1600cc. According to the evolution model, this data would imply that Neanderthal-type peoples evolved

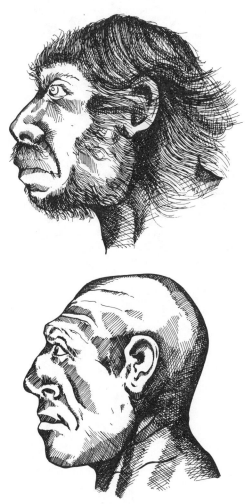

Figure 45 The two illustrations of Neanderthal Man from fossil skull data has changed with man's perception on how these humans fit into the evolution model.

57

from modern man. Or perhaps the human brain is actually shrinking! A better explanation is that the variation in brain size seen in human fossils is consistent with the variation predicted by genetics and the creation model.

What You Believe Does Matter

A model is important in that it affects the way people live and plan for the future. When a society believes that evolution is true, evolution becomes the model which people use to organize society and prepare for the future. Combining belief in evolution with powerful social institutions such as business (for example, the **abortion** industry) and politics (for example, **racism**) is especially dangerous and far reaching. As an extreme example, there are people who believe that human development must be controlled. They argue that during human evolution from ape to man, weaker individuals either died, or in other ways were prevented from having offspring. Their weaknesses could not be passed to future generations. Today we have developed technology and social welfare systems which prevent this "natural selection" from occurring. Therefore, society must control the lives of inferior people to prevent endangering the quality of life for future generations. (Crow, 1966; Osborn, 1960) Studies concerning the control of human heredity are called **eugenics**. These studies are directly related to evolution. The eugenics-evolution connection could be a force to control the lives of people that might be considered inferior or genetically weaker.

The creation model, on the other hand, provides for the sanctity of human life, and those adhering to this model strongly object to genetic manipulations. Truly, what you believe about human origins does matter!

Less than forty years ago a political group called Nazis gained control of Germany. Believing in evolution, they wanted to produce a "Super Race." To do this they developed programs to separate this "Super Race" from the people they considered inferior. Many people considered inferior were isolated in work camps and many more were horribly murdered in death camps. What you believe about human origins does matter!

Figure 46 One issue debated about abortion is whether or not a baby is a human being from the momemt of conception. A false concept often used to justify abortion is that the baby goes through evolutionary stages: fish stage, an amphibian stage, a reptile stage, and so on. This is not true. (see Chapter 3, pages 28–29)

"Oldest Man" or Fossil Ape?

John Reader has given us a rare and revealing insight into how the "human-ness" of ape fossils can be arrived at by scientists. The story is about a fossil skull known as "1470" which was popularly portrayed by the media to the world as the "Oldest Man."

In September 1973, Richard Leakey called a meeting in Nairobi to discuss the formal scientific description of the discovery. Attending the meeting were the equally well-known anthropologists Bernard Wood, Alan Walker and Michael Day. According to Reader, the meeting didn't go smoothly. Alan Walker was the scientist most involved with the reconstruction of the fossil skull fragments. Dr. Walker argued for the australopithecine affinities of the skull. However, Leakey and Wood argued that skull "1470" indicated a "large brain" and, therefore, must be *Homo.*

"In the ensuing debate Walker detailed the australopithecine affinities and insisted that they merited nomenclature acknowledgment regardless of brain size. Furthermore, the skull had been distorted during fossilization, he said, and the configuration of the right side of the vault was squashed into a deceptively *Homo*-like form. But for Leakey and Wood, brain size was all that counted. As the debate warmed to argument with no concession from either side, Walker resorted to persuasion of a more personal nature. If the published description of 1470 was to include *Homo* attribution, he said, then his name must be removed from the paper. This was no mean threat, given Walker's academic standing and contributions to the field , but it did not bring the capitulation he sought. Quite the contrary in fact, for the threat drew from across the table the flippant and injudicious remark that his withdrawal might be welcomed. At this Walker picked up the fossil and left the room.

After his departure, Richard Leakey decided to resolve the conflict in the following manner: if Walker stayed away then 1470 would be attributed to *Homo* without qualification; if he returned then Walker's views would be accommodated, to some degree, in the paper. Walker did return and the paper was published with his name beneath the innocuous title: "New Hominids from East Rudolf, Kenya." (p. 204)

Reference: *Missing Links: The Hunt for Earliest Man*, John Reader, (1981).

Evolution of Man? Australian Fossils Say No!

Human fossil skeletons discovered in Australia provide examples of fossils that go against the evolutionary models of modern man evolving from *Homo erectus*. Neanderthal people should have lived tens of thousands of years before people with skeleton types like those of modern humans. It is significant that in some parts of the world this hypothetical sequence is reversed. Australia is one region where modern skeleton types apparently lived before *Homo erectus* . In fact, if one has confidence in the evolutionary biased radiometric dating system, the "modern"-type fossils were from people who lived many thousands of years before the *Homo erectus.*

These "modern"-type fossil skeletons were found in eroding sand dunes along the edge of a desert lakebed called Lake Mungo. When these people were alive, lakes extended into what is now the semiarid area west of New South Wales. Animal remains discovered around the ancient campfires revealed that these people ate fish caught from the lake and killed small animals such as wallabies. As some Australian Aborigines do today, emu eggs were collected for food by the Lake Mungo people.

Another archaeological site on the Victorian border with New South Wales is known as Kow Swamp. Here were discovered the fossilized bones of about 60 people. These people are believed to have lived long after the Lake Mungo people had died and the lakes around which they lived had dried. The important feature of these human fossils is that the skulls are even more "primitive" in appearance than the *Homo erectus* skull used as the classification standard. Some evolutionists have suggested that the Kow Swamp people evolved from the modern-type Mungo people. Others believe that these *Homo erectus* types must be a different "race" which mixed with the "race" of modern-type people to produce today's Australian Aborigine.

Whether the Kow Swamp fossils represent human genetic variation or an evolutionary ancestor to modern man can have an effect on how a person views the Australian Aborigine. What are some of these views resulting from the creation and evolutionary models of human origins?

Kow Swamp Burial

Lake Mungo Burial

CHAPTER NINE

Time and Earth History

Time and earth history is perhaps the most debated area in the whole question of the origin of life. Everyone seems to have his own idea about how old the earth is. In a real sense time is only vitally important to the evolutionist. Evolution needs lots of time for life to have arrived on the scene. Short time periods would seriously undermine the theory. On the other hand, the creationist says that an intelligent designer outside this universe made the earth and the life on it. Whether this life was created in six days or six billion years is insignificant to the scientific view of creation. The issue of time, then, is critical for the evolutionist — not for the creationist.

I will touch on a variety of important time measurements such as **radiometric dating**, **oil well pressure**, **decay of earth's magnetic field**, **polonium halos**, and others. Be critical of your own bias and ask questions about the standard scenario of five billion years. How does one select a time clock? How does one know if the "clock" is giving the correct time?

Figure 47 Time Clocks. Evolution depends upon billions of years of time because the various processes take a long time. There are certain kinds of time clocks that are believed to tell the age of the earth and give scientists clues for great amounts of time.

Frequently we hear that anyone can see this is an ancient, billion-year-old earth that we are living upon. One resource for teachers says that this whole question has no relevance because we *know* the earth is 4.5 billion years old. Have you asked the question, what would a young earth look like?

When we look at many geologic formations we see evidences of catastrophe. But if we see a valley with a small stream running through it, can we assume this is evidence for long time periods? To answer these questions geologists, like Dr. Steve Austin, have been carrying out valuable studies of the catastrophe at Mount St. Helens and its effect in producing geological formations. We can view his work with great interest because it also seems to relate directly to questions we have about other geological structures on planet earth.

> The explosion of Mount St. Helens in Washington State on May 18, 1980 was initiated by an earthquake and rockslide involving one half cubic mile of rock. As the summit and north slope slid off the volcano that morning, pressure was released inside the volcano where super-hot liquid water immediately flashed into steam. The northward-directed steam explosion released energy equivalent to 20 million tons of TNT which toppled 150 square miles of forest in six minutes. In Spirit Lake north of the volcano, an enormous water wave initiated by one eighth cubic mile of rock slide debris stripped trees from slopes as much as 850 feet above the pre-eruption water level. The total energy output on May 18 was equivalent to 400 million tons of TNT, or approximately 20,000 Hiroshima-size atomic bombs. (Austin, 1986)

Dr. Austin points out that up to 600 feet of strata (layers of rock) were formed since 1980 at Mount St. Helens, and this from one of the smaller catastrophes in earth history! He states that Mount St. Helens provides a rare opportunity to study transient geologic processes which produced, within a few months, changes which geologists might otherwise assume might require many thousands, or even millions, of years.

Figure 48 A volcano is but one small example of the kind of catastrophes that leave evidence in the geologic record.

How Are Rocks Dated?

Hundreds of thousands and even billions of years are claimed for radiometric dating techniques. No one was there to verify the initial amount of the radioactive mineral. No one was there to watch the radiometric decay taking place. We assume that it must have been constant and unchanging from what we can measure today.

Consequently, it is very important that we consider carefully the assumptions that have to be made. If our assumptions are wrong then our results will be wrong.

Assumptions for Radiometric Dating

1. **The system must have been a closed system**. By this we mean that the system cannot be altered by contamination from the outside or loose material from inside the system migrating to the outside. It must be constant.

2. **The system must have a starting point – a beginning**. For example, there would have to be just uranium and no lead in the beginning of the Uranium 238 and Lead 206 series.

3. **The rate of decay must always have been the same.** If decay rates of the material vary, the resulting measured time cannot be trusted. Is there such a thing, in nature, as a closed system? Is it possible to know if the system started at a zero set point? Are we sure about the rate of decay way back in history? In addressing these questions, we will deal with generalized information and then direct you to more detailed sources if you choose to explore further.

Dating techniques are all based on certain assumptions. The **true age** depends entirely upon the validity of the assumptions. Let's look at these assumptions and see how they stand up to the test.

Uranium Dating Method

First, the uranium method is really a whole family of decay methods and not just one. The method depends upon **uranium** and its sister element **thorium** going through long decay chains until finally the stable element, Lead 206 is reached. In this process, alpha particles are given off. In fact, eight alpha particles are given off from each Uranium 238 (^{238}U) atom that decays to lead 206 (^{206}Pb). Looking at the diagram (figure 49) you can get some idea of what happens.

In evaluating the uranium dating method, the fact is that uranium minerals always exist in open systems. This causes us to immediately question the validity of the method. One of the chief authorities on radioactive dating, Dr. Henry Faul, said:

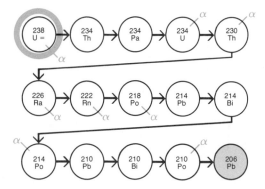

Figure 49 This diagram shows how uranium (U-238) decays to lead (Pb-206). Creation scientists show this clock can give young ages more logically than old ages. They also show that these clocks may not always decay at a constant rate.

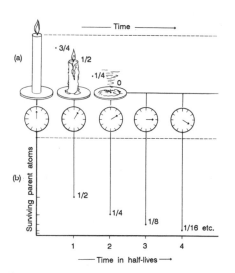

Figure 50 (a) Uniform straight line depletion of most everyday processes. (b) By contrast, the radioactive decay curve approaches zero line asymptotically. The end of one half-life interval is the beginning of a new one.

Uranium and lead both migrate (in shales) in geologic time, and detailed analyses have shown that useful ages cannot be obtained with them. Similar difficulties prevail with pitchblend veins. Here again widely diverging ages can be measured on samples from the same spot. (Faul, 1966)

Remember that one criterion for valid dates is a **closed system**. We find that all of the age dating methods (rubidium-strontium, uranium-lead and potassium argon) can give different and widely varying dates. The lead ages, for example, have given consistently older dates then the others. This very fact encouraged Dr. Leon T. Silvers of the California Institute of Technology to do an experiment that showed lead could vaporize and move out of the sample. (Driscoll, 1972) A scientist could calculate a "low" age for the sample from which the lead escaped and a "high" age for the sample where the lead migrated to.

Potassium-Argon Dating Method

This method is the most popular method for dating rocks, and is used extensively to give general dates for fossils found in some rock formations. Potassium minerals are found in igneous and in some sedimentary rocks, therefore the Potassium-Argon (K-Ar) method isn't as restricted in its use as is uranium. Here is what happens: **Potassium^{40}** decays into **Argon^{40}** when its molecules capture an electron. **Potassium^{40}** has a half life of 1.3 billion years. In other words, one half of the **Potassium^{40}** has decayed in 1.3 billion years. The diagram below will give you some idea of what this process is like.

The K-Ar system behaves a little differently from the uranium series above. An amount of ^{40}K decay (11%) will become ^{40}Ar, instead of ^{40}Ca. The ratio of potassium to argon is considered in the calculations for the age factor of a particular rock. But again, we find that the ideal does not exist in nature. We don't know for sure how much potassium and argon were **originally** there in the sample. Furthermore, we don't have any way of knowing that the sample was not contaminated in nature or exposed to migration.

The Question of Helium

Alpha particles, which are actually helium nuclei and are positive (+) are given off by the decay of the Earth's radioactive isotopes. Eventually, the helium migrates out of rocks in the earth's crust and ends up in the atmosphere as helium gas.

Here these particles of helium gas reach their upper limits in the atmosphere where a few molecules escape (figure 51). There is continuing research going on concerning the helium inventory in the atmosphere and one of the scientists investigating this problem is Dr. Larry Vardiman of the Institute for Creation Research. Even though helium in the atmosphere does not date the rocks, it does tell us something about the earth's age.

> If the earth was billions of years old, the radioactive production of helium in the earth's crust should have added a large quantity of helium to its atmosphere. Current diffusion models all indicate that helium escapes to space from the atmosphere at a rate much less than its production rate. The low concentration of helium actually measured would suggest that the earth's atmosphere must be quite young. (Vardiman, 1986)

If the low concentration of helium in the atmosphere is valid then we must consider this as a positive evidence for a young earth.

Radiocarbon Dating

Most of you will recognize this as "carbon" dating. This type of dating is confined to things that were once alive. All living things are made of molecules containing carbon. Carbon also exists in the gases of our atmosphere, such as CO, CO_2, etc. In addition to this all of our food is made up of carbon in some form. For example, our fats, sugars, and many other molecules are carbon based (organic). We can understand why ^{14}C, **radioactive carbon**, as well as **non-radioactive carbon**, ^{12}C, can become part of all parts of a living plant or animal. ^{14}C gets into a living system as some ratio of $^{14}C/^{12}C$. ^{14}C and ^{12}C are continually taken into the living system. When the organism dies then the $^{14}C/^{12}C$ ratio changes as the radioactive ^{14}C begins to decay to the stable ^{14}N. We can measure the ratio of $^{14}C/^{12}C$ in living things and also measure the $^{14}C/^{12}C$ ratio in a dead organism. Knowing the decay rate of ^{14}C, (a half-life of 5730 years) we can tell how old the organism is. Unfortunately, as with the other radiometric dating methods, there are assumptions and these must be considered.

Where does the radioactive carbon come from? Cosmic rays strike air particles (78% nitrogen, 21% oxygen, 1% other gases) and knock the **neutrons** out of them. Colliding neutrons cause protons to leave the nitrogen atoms (^{14}N) which causes them to change to ^{14}C — radioactive carbon. Any ^{14}C atom can attach to oxygen to make carbon dioxide (CO_2) which is one of the gases that we breathe. Therefore, they thoroughly permeate living systems. Once the living organism dies then the ^{14}C is no longer replenished and the $^{14}C/^{12}C$ ratio changes as the radioactive carbon decays to non-radioactive ^{14}N.

When ^{14}C was first suggested as a dating method, it was assumed that the earth was billions of years old. If the world were this old, the amount of ^{14}C produced in the atmosphere would have equilibrated — it would have become a stable, predictable amount. But what if the earth is young? The amount of radioactive carbon in the past would be less than it is today. This error would mean that ^{14}C dating methods would be more accurate for dating recently dead materials (only a few thousand years dead), but would give ages with increasingly greater error the farther back in history from which the sample comes. In other words, there is much work yet to be done to calibrate and validate this method before it becomes an accurate measure of time.

Polonium Halos

Polonium halos are perhaps the most significant evidence for a young earth. Dr. Robert Gentry's landmark study of polonium halos is published, in detail, in the book *Creation's Tiny Mystery* (1986).

A little history on the "radio halos" of polonium will help you understand the significance of these microscopic marks in the earth's first or "oldest" rocks. If you refer to the uranium decay series, you will see elements (isotopes) ^{218}Po, ^{214}Po and ^{210}Po.

All of these isotopes are radioactive and give off radioactive particles. Now the interesting part is that the Po isotopes all have short **half lives** (the time needed for half of the Po to decay). ^{218}Po has a half life of **3 minutes**, ^{214}Po a half life of **164 microseconds**, and ^{210}Po has a half life of **138.4 days**

What does this mean? First of all, the halos (halos formed by radioactive decay) are from 218, 214, and 210 isotopes of polonium that has no connection with uranium. In other words, *they exist by themselves without any uranium parent present!* When they are found in the oldest rocks (granites and biotites) one has to imagine that the rocks must have solidified very quickly in order to allow the decay of the radioactive particles to make these distinct marks. If the first rocks cooled slowly (as all evolution models demand) then in molten rock there could be no marks from the decay. The decaying particles can only make their marks after the molten rocks have solidified. Gentry concludes that this scientific evidence establishes that the age of the earth's oldest rocks cannot be more than thousands of years. Po halos in our oldest rocks give strong evidence as a **limiting chronometer**, that is, a time clock which can set an upward limit on the age of the earth's crust. But here is the interesting part: if the evolutionist refuses to accept that the oldest rocks were instantly created, as evidenced by the polonium halos, then he must accept that decay rates are **not constant**! This admission then puts the third assumption of radiometric dating (constant decay rates) into disarray.

The radiometric dating methods discussed are but one type of time clock. Below are some other time clocks,

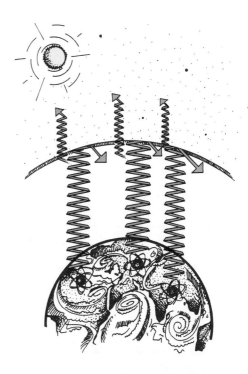

Figure 51 If the earth were billions of years old we would expect much more helium in the atmosphere. The low concentration of helium in the atmosphere indicates that the earth is young.

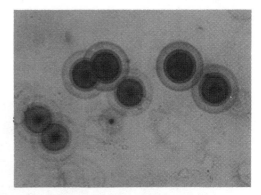

Figure 52 "Radio halos," formed by radioactive decay, strongly suggest that the age of earth's oldest rocks cannot be more than a few thousand years old.

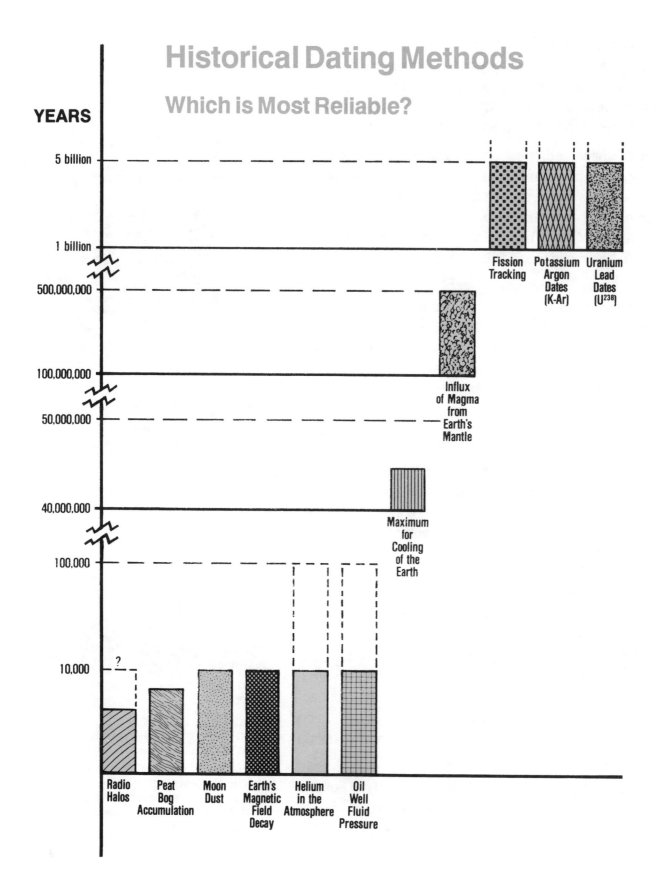

Historical Dating Methods

Which is Most Reliable?

YEARS

5 billion

1 billion

500,000,000

100,000,000

50,000,000

40,000,000

100,000

10,000

?

Radio Halos

Peat Bog Accumulation

Moon Dust

Earth's Magnetic Field Decay

Helium in the Atmosphere

Oil Well Fluid Pressure

Maximum for Cooling of the Earth

Influx of Magma from Earth's Mantle

Fission Tracking

Potassium Argon Dates (K-Ar)

Uranium Lead Dates (U^{238})

which in many cases require fewer assumptions that the radiometric methods discussed above.

Oil Well Fluid Pressures

"Oil well fluid pressures" are a limiting chronometer that speaks to young earth ages. We are told by geophysicists that the existing pressures found in these deep wells would cause the oil to seep out of the formations in which it was trapped if the true age of the earth were billions of years. (Cook, 1968)

Eroding of the continents

Eroding of the continents seems to indicate that if this process had been going on for millions of years (14 million is the present calculation) then all of the present continents, eroding at present rates, would have been eroded to sea level. (Nevins, 1978)

Decay of the Earth's Magnetic Field

Decay of the Earth's Magnetic Field can be used as a limiting chronometer. Dr. Thomas Barnes has calculated the present strength of the earth's magnetic field and extrapolated (calculated) backwards. He has concluded that the strength of the magnetic field a million years ago would have been so great that it would have torn the earth apart. This information limits what the age of the earth could be. He estimates that it took thousands of years, rather than billions of years, for the earth's magnet to have decayed to its present strength. Now, there are no questions about the existence of the magnetic field but there are many questions about how it got here and whether it has undergone numerous reversals. At this point in time Dr. Barnes has firmly refuted the various challenges that have been offered by some against this strong evidence. (Barnes, 1983)

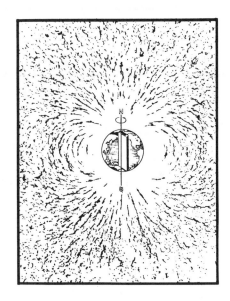

Figure 53 This shows how the earth acts as a huge magnet.

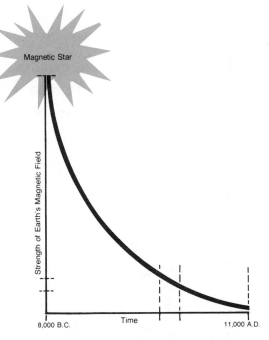

Figure 54 This exponential decay rate comes from 135 years of observation of the earth's magnetic field. It is interesting to note that radioactive decay rates have not been observed for this length of time.

69

Dating the World's Oldest "Man"

In Chapter 7, we looked at the fossil ape skull known as "1470." This fossil was touted by the popular media as the "Oldest Man." It was dated using the potassium-argon method. John Reader in his book *Missing Links* (1981) gives us a glimpse of the method in use.

"The Worldwide admiration and congratulations that greeted the twenty-eight-year-old Richard Leaky and his two-and-a-halfmillion-year-old 1470 (number given to the fossil skull) were subsequently marred by just one thing—an authoritative suggestion that the skull was not as old as Leaky claimed . . .

Fitch and Miller's tests on the first samples . . . that Leakey sent to Cambridge actually gave an average age of 221 million years. Such an age was impossible—so Leakey sent more samples. From these the scientists selected crystals that seemed fresher than others and produced an age of 2.4 million years. . . They subsequently tested many more samples (including some they had collected themselves) and their results range from a minimum of 290,000 years to a maximum of 19.5 million." (pages 205-206)

Textbooks often refer to radiometric dating methods as **absolute** dating techniques. What is the significance of such radiometric dating methods? What method gives you the greatest confidence as a measure of time in the past?

CHAPTER TEN

Creation or Evolution?

Popular scientific opinion, rather than scientific objectivity, has often prevailed in the past and, unfortunately, prevails today. History has shown that narrow-mindedness and dishonesty in inquiry have hurt the scientific enterprise. Good education in science or any other area requires an objective comparison of ideas. To present only one view in the classroom as established fact when it clearly is not, is not education. This approach is indoctrination in its grossest form.

My attempt has been to give you "starter" information on the origin of life. In this way you can carry out your own expanded research so that you can continue to develop models compatible with the evidence. Although there was no one on the scene to witness the beginning, we do have the mentality, as humans, to draw inferences from the scientific data that we can observe.

Some have argued that creation science cannot be true science because creation depends upon a Creator and science cannot deal with a Creator. Now you have the information to handle this bias. Either a Creator made the matter and energy of the universe or the universe made itself. Again, test the evidence. Do we see evidence of continuous development from molecules to man anywhere? We don't, even though some of the greatest intellects of our age have desperately sought such evidence. We must conclude that a great measure of faith is also required to fill all of the major gaps and assumptions that evolutionists must make within their model.

The atheist or secular humanist can't help but to assume evolution. He has rejected the idea of a God of creation. Agnostics and many liberal theologians are not really concerned about the outcome of the scientific evidence, because they are biased against the Bible and the Creator.

And the Biblical creationist is certain, based upon his belief in Scripture, that God created man as man, dog as dog, mouse as mouse — all life that we see. All of these are faith beliefs that science has no access to. The scientific creationist must do good science and report honestly even though he has, as his foundation, a belief in God as a creator of all things. The same thing is true for the atheist scientist. Faith characterizes all scientists engaged in scientific endeavor. None of the men or women engaged in science are without bias; however, they can all practice good science and many do.

Whether creationist or evolutionist, when doing science, we must do science in the best manner possible. To restate the theme expressed by The National Academy of Science resolution mentioned in the Introduction . . . *Scientific research should be conducted with intellectual freedom.*

Selected Bibliography

Books identified with an (*) are creation science oriented in content, and most are available through the Institute for Creation Research, 800/628-7640 (or at www.icr.org/store).

Abel, O., *Palaeobiologic und Stammesqeschichte* (Jena, 1971), p. 285-286, 294.

Ager, D., "The Nature of the Fossil Record," *Proceedings of the British Geological Society*, March 5, 1976.

———, "Fossil Frustrations," *New Scientist*, November 10, 1983, p. 425.

* Austin, S., "Mount St. Helens and Catstrophism," *Proceedings of the First International Conference on Creationism,* Vol. 1 (Pittsburgh, PA:Creation Fellowship, 1986).

Axelrod, D., *Science*, 128.7, 1958.

* Barnes, T., *Origin and Destiny of the Earth's Magnetic Field* (San Diego, CA: Institute for Creation Research, 1983).

Bellwood, P., *Man's Conquest of the Pacific* (New York, NY: Oxford University Press, 1979).

Birdsell, J., *Human Evolution* (Chicago, IL: Rand McNally & Co., 1975), p. 170.

Bleibtreu, H. *Britannica Book of the Year* (Chicago, IL: Encyclopedia Britannica, Inc., 1985), p. 163.

* Bliss, R., *Good Science for Home and Christian Schools* (San Diego, CA: Institute for Creation Research, 1987).

* Bliss, R., and G. Parker, *Origin of Life* (Green Forest, AR: Master Books, 1979).

* ———, *Fossils: Key to the Present* (Green Forest, AR: Master Books, 1980).

* Cook, M. *Prehistory and Earth Models* (London: Max Parrish, 1966), p. 254-259).

Cox, B., "Mysteries of Early Dinosaur Evolution," *Nature*, 264:314, 1976.

Cherfas, J., "Leakey Changes His Mind about Man's Age," *New Scientist*, 1982, p. 125.

Crow, J., "The Quality of People: Human Evolutionary Changes," *Bioscience*, 16:863-867, 1966.

Dayhoff, M., *Atlas of Protein Sequence and Structure* (MA: National Biomedical Research Foundation, 1972).

de Beer, Sir Gavin, "Homology: An Unsolved Problem, *Oxford Biology Readers* (New York, NY: Oxford University Press, 1971).

* Denton, M., *Evolution: A Theory in Crisis* (London: Burnett Books Limited, 1985).

Driscoll, E., "Dating of Moon Samples: Pitfalls and Paradoxes," *Science News*, 101:12, 1972.

———, "Evolution: Concepts and Consequences," *Scientific American*, January 1975.

Faul, H., *Ages of the Rocks, Planets, and Stars* (New York, NY: McGraw Hill, 1966), p. 61.

Feduccia, A., and H. Tondoff, "Asymmetric Vanes Indicate Aerodynamic Function," *Science*, 203:1020, 1979.

Fox, J., "DNA Replication: More Complicated than Ever," *Chemical and Engineering News*, June 19, 1978.

* Gentry, R., *Creation's Tiny Mystery* (Knoxville, TN: Earth Science Associates, 1986).

* Gish, D., *Evolution: Challenge of the Fossil Record* (Green Forest, AR: Master Books, 1985).

* ———, *Dinosaurs — Those Terrible Lizards* (Green Forest, AR: Master Books, 1976).

Gould, S., "Piltdown Revisited," *Natural History*, 88:86, 1979.

Gould, S., and M. Eldredge, "Punctuated Equilibria: the Tempo and Mode of Evolution Reconsidered," *Paleobiolgy*, 6:115-151, 1977.

Gribbin, J., and J. Cherfas, "Descent of Man — or Ascent of Ape?" *New Scientist*, 592-595, September 3, 1981.

* Ham K., *The Lie: Evolution* (Green Forest, AR: Master Books, 1987).

Hardin, G., "Nature and Man's Fate," 1960, p. 225-226.

Hause, M., ed., *The Origin of Major Invertebrate Groups,* Special Vol. 12 (New York, NY: Academic Press, 1979).

Hitching, F., *The Neck of the Giraffe* (New York, NY: Ticknor and Fields, 1982).

Howell, W., *Mankind in the Making* (New York, NY: Doubleday, 1967).

Hoyle, F., and C. Wickramasinghe, *Evolution from Outer Space* (New York, NY: Simon and Schuster, 1981).

———, *Humanist Manifesto*, Vol. 1 and 2 (Buffalo, NY: Prometheus Books, 1977).

Ivanhoe, F., "Was Virchow Right about Neanderthal?" *Nature*, 227:577, 1970.

Jeletzsky, J., "Paleontology as a Basis of Practical Chronoloy," *Bulletin of American Association of Petroleum Geologists*, 40:685, 1956.

Johanson, D., and M. Edey, *Lucy: the Beginning of Mankind* (New York, NY: Simon and Schuster, 1981).

Jones, R., "The Fifth Continent: Problems Concerning the Human Colonization of Australia," *Annual Review of Anthropology*, 8:445-466, 1979.

Kelso, A., *Physical Anthropology* (New York, NY: Lippincott, 1974).

Kerkut, G., *Implications of Evolution* (London: Pergamon Press, 1960).

Kusinitz, M., *Science World*, February 4, 1983, p. 12-19.

Leakey, R., *The Making of Mankind* (London: Michael Joseph, Ltd., 1981), chapters 13 and 14.

Levine, J., "New Ideas about the Early Atmosphere," NASA Special Report No. 225, Langley Research Center, August 11, 1983.

Lewin, R., "How Did Humans Evolve Big Brains?" *Science*, 216:840-841, 1982.

Mayer, E., *The Growth of Biological Thought* (Cambridge, MA: Belknap Press, 1982).

* Mehlert, W., "The Australopithecines and (Alleged) Early Man," *Creation Research Science Quarterly*, 17:23-25, 1980.

Miller, S., and L. Orgel, *The Origins of Life on the Earth* (New York, NY: Prentice Hall, 1974).

* Morris, H., *Scientific Creationism* (Green Forest, AR: Master Books, 1985).

* ———, *Men of Science, Men of God* (Green Forest, AR: Master Books, 1988), p. 49.

* ———, *History of Modern Creationism* (Green Forest, AR: Master Books, 1984).

* Morris, H., and G. Parker, *What Is Creation Science?* (Green Forest, AR: Master Books, 1987).

———, "An Affirmation of Freedom of Inquiry and Expression," National Academy of Sciences Resolution, April 1976.

* Nevins, S., "Evolution: the Oceans Say No!" *Acts & Facts*, Impact Series No. 8, October 1978.

Olson, E., and J. Robinson, *Concepts of Evolution* (Merrill, 1975).

Olson, S., "Flight Capability and the Pectoral Girdle of Archaeopteryx," *Nature*, 178:247, 1979.

Ommaney, F., *The Fishes* (New York, NY: Time Life, Inc., 1960), p. 60.

O'Rouke, "Pragmatism Versus Materialism in Stratigraphy," *American Journal of Science*, January 1976, p. 53.

Osborn, F., "A Return to the Principles of Natural Selection," *Eugenics Quarterly*, 7:204-211, 1960.

Oxnard, C., and F. Lisowske, *American Journal of Physical Anthroplogy*, 52:116, 1980.

Pilbeam, D., *The Evolution of Man* (New York, NY: Funk and Wagnalls, 1970).

Ransom, J., *Fossils in America* (New York, NY: Harper and Row, 1964), p. 43.

Reader, J., *Missing Links* (London: Collins, 1981), p. 9, 73-110).

Rensberger, B., "Ancestors: a Family Album," *Science Digest*, 89:34-43, 1981.

Rensch, B., *Evolution from the Species Level* (New York, NY: Columbia University Press, 1959), p. 166.

Romer, A., *Vertebrate Paleontology* (Chicago, IL: University of Chicago Press, 1966), p. 12.

Saritch, V., "Some Thoughts on Evolution and Creation," Bakersfield debate, May 10, 1986.

Simmons, E., "Keynote address," Ann. N.Y. Acad. Sci., 167:319, 1969.

Spieker, E., "Mountain-Building Chronology and the Nature of the Geologic Time Scale," *Bulletin of American Association of Petroleum Geologists*, 40:1805, 1956.

Stanley, S., *Macroevolution Patterns and Processes* (San Diego, CA: Freeman, 1979).

———, *The New Evolutionary Timetable* (New York, NY: Basic Books, 1981).

* Sunderland, L., *Darwin's Enigma: Fossils and Other Problems* (Green Forest, AR: Master Books, 1987).

Susman, R., and J. Tern, "Functional Morphology of Homo Habilis," *Science*, September 3, 1982.

* Taylor, I., *In the Minds of Men* (Toronto: TFE Publishing, 1987).

Trinkaus, E., and W. Howells, "The Neanderthals," *Scientific American*, 241:118, 1979.

* Unfred, D., and J. Mackay, "Little-Known Facts about Dead Apes," *Creation (Ex Nihilo) Magazine*, 8:15-20, 1986.

* Vardiman, L., "The Age of the Earth's Atmosphere Estimated by its Helium," *Proceedings of the First International Conference on Creationism*, Vol. 2 (Pittsburgh, PA: Creation Science Fellowship, 1986).

Wallace, R., et al, *Biology: the Science of Life* (Scott Foresman, 1986).

Zuckerman, S., *Beyond the Ivory Tower* (New York, NY: Taplinger Publishing Co., 1970), p. 75-94.

VIDEOS:

Origins: *Creation or Evolution*, Master Books, Inc., P.O. Box 727, Green Forest, AR 72638.

Answers in Genesis, a complete creation seminar on video, Ken Ham and Gary Parker, Master Books, Inc., P.O. Box 727, Green Forest, AR 72638.

In his book *Men of Science, Men of God,* Dr. Henry Morris reviews the scientific achievement of over 100 famous scientists — scientists who also believe in the Creator. Here, briefly, is the story of one of these men.

In Psalm 8:8 is a small piece of information "And whatsoever passeth through the paths of the sea."

U.S. Naval Commander Matthew Maury read this in his Bible and came to the conclusion that the sea must have "paths." He began a scientific study in 1841 on the basis of his faith that the Bible is true, even when it speaks in areas of science. Maury launched an investigation that lasted for 20 years. During this time he charted the great currents of the sea, such as the Gulf Stream and the Labrador currents.

From his study, which began with Psalm 8 in the Bible, he became known as the "Pathfinder of the Seas." The work which Maury started eventually became the U.S. Naval Observatory. Good science doesn't eliminate any possibility toward the goal for truth. If statements in the Bible lead to scientific discovery, then why not explore the historical record to see what other scientific truths can be discovered?

The most thoroughly tested history book ever written is the Bible. It has proven to be an accurate and reliable record — even in ages when popular opinion was against it. But beyond its historical accuracy is the truth that it tells about our Creator, the One who made us, the One who knows us.

$7.95 • paperback
108 pages
Age level: 12–adult
ISBN: 0-89051-080-6

For orders or information:

800/628-7640 or www.icr.org/store

Institute for Creation Research

P.O. Box 2667 • El Cajon • CA 92021

"Science, Scripture, & Salvation" Extended Versions

For More Information

This list of recommended books is just a hint of what is available about the exciting story of creation and creation science from the **Institute for Creation Research.**

Men of Science, Men of God
by Dr. Henry Morris

Over 100 exciting mini-biographies of great scientists of the past and present who believed in the Creator. Many of our "founding fathers" in the fields of modern science were creationists.

ISBN: 1-89051-080-6
Paperback • **$7.95**

The Lie: Evolution
by Ken Ham

At an ever-accelerating pace, society is putting its stamp of approval on practices that just several decades ago were not only frowned upon, but were outright illegal. Ken Ham gets to the bottom of the problem in this book, showing how we have been simply fighting the symptoms and overlooking the root cause.

ISBN: 0-89051-158-6
Paperback • **$9.95**

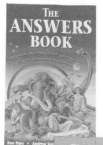

The Answers Book
by Ken Ham, Andrew Snelling, and Carl Wieland

This fascinating book addresses the most common questions that Christians and non-Christians alike ask regarding creation/evolution and Genesis: Where did Cain get his wife? What about contiental drift; the origin of the nations, dinosaurs, the Gap Theory, carbon dating, etc.? The 12 most-asked questions are answered in an easy to understand manner with helpful illustrations.

ISBN: 0-89051-161-6
Paperback • **$11.95**

Creation: Facts of Life
by Gary Parker

Dr. Gary Parker, leading creation scientist and speaker, presents the classic arguments for evolution used in public schools, universities, and the media, refuting them in an entertaining and easy-to-read style. Once an evolutionist himself, Dr. Parker is well-qualified to expose these arguments. A must for students and teachers alike.

ISBN: 0-89051-200-0
Paperback • **$10.95**

Noah's Ark and the Ararat Adventure
Dr. John D. Morris

This Gold Medallion finalist and truly pleasurable read includes Dr. John Morris's latest clues and information as to whether or not the ark is possibly on Mt. Ararat. He discusses the stone formation in Turkey that many believe is the ark. Illustrations and Dr. Morris's personal photos enhance the quality and the evidence of the ark within this book for anyone, ages 6 to adult.

ISBN: 0-89051-166-7
Hardcover • **$13.95**

 Orders or information: 800/628-7640 or www.icr.org/store